CHEMISTRY RESEARCH AND APPLICATIONS

PYRROLE

SYNTHESIS AND APPLICATIONS

CHEMISTRY RESEARCH AND APPLICATIONS

Additional books and e-books in this series can be found on Nova's website under the Series tab.

CHEMISTRY RESEARCH AND APPLICATIONS

PYRROLE

SYNTHESIS AND APPLICATIONS

COLIN WELCH
EDITOR

Copyright © 2020 by Nova Science Publishers, Inc.

All rights reserved. No part of this book may be reproduced, stored in a retrieval system or transmitted in any form or by any means: electronic, electrostatic, magnetic, tape, mechanical photocopying, recording or otherwise without the written permission of the Publisher.

We have partnered with Copyright Clearance Center to make it easy for you to obtain permissions to reuse content from this publication. Simply navigate to this publication's page on Nova's website and locate the "Get Permission" button below the title description. This button is linked directly to the title's permission page on copyright.com. Alternatively, you can visit copyright.com and search by title, ISBN, or ISSN.

For further questions about using the service on copyright.com, please contact:
Copyright Clearance Center
Phone: +1-(978) 750-8400   Fax: +1-(978) 750-4470   E-mail: info@copyright.com.

NOTICE TO THE READER

The Publisher has taken reasonable care in the preparation of this book, but makes no expressed or implied warranty of any kind and assumes no responsibility for any errors or omissions. No liability is assumed for incidental or consequential damages in connection with or arising out of information contained in this book. The Publisher shall not be liable for any special, consequential, or exemplary damages resulting, in whole or in part, from the readers' use of, or reliance upon, this material. Any parts of this book based on government reports are so indicated and copyright is claimed for those parts to the extent applicable to compilations of such works.

Independent verification should be sought for any data, advice or recommendations contained in this book. In addition, no responsibility is assumed by the Publisher for any injury and/or damage to persons or property arising from any methods, products, instructions, ideas or otherwise contained in this publication.

This publication is designed to provide accurate and authoritative information with regard to the subject matter covered herein. It is sold with the clear understanding that the Publisher is not engaged in rendering legal or any other professional services. If legal or any other expert assistance is required, the services of a competent person should be sought. FROM A DECLARATION OF PARTICIPANTS JOINTLY ADOPTED BY A COMMITTEE OF THE AMERICAN BAR ASSOCIATION AND A COMMITTEE OF PUBLISHERS.

Additional color graphics may be available in the e-book version of this book.

Library of Congress Cataloging-in-Publication Data

ISBN: 978-1-53617-137-2

Published by Nova Science Publishers, Inc. † New York

CONTENTS

Preface		vii
Chapter 1	Polypyrrole Based Fabrics: Synthesis and Applications (Review) *Smita Deogaonkar-Baride, Narendra V. Bhat, Padma Vankar and Anjan K. Mukhopadhyay*	1
Chapter 2	Recent Research Advances in the Aqueous Phase Synthesis of Pyrroles *Venkata Durga Nageswar Yadavalli, Swapna Kokkirala and Venkanna Avudoddi*	45
Chapter 3	UV and Visible Light Photoinduced Polymerization of Pyrrole/Methacrylate *Claudia I. Vallo and Silvana V. Asmussen*	71
Index		99
Related Nova Publications		105

PREFACE

Pyrrole: Synthesis and Applications provides an overview of polypyrrole synthesis by different methods such as chemical polymerization, lecetro-polymerization, photo-initiated polymerization and γ irradiated polymerization.

Pyrrole derivatives play a crucial role in organic chemistry, medicinal chemistry, and heterocyclic chemistry. Pyrrole scaffold is extensively used in the synthesis of drug molecules with various pharmacological properties, as well as in material sciences.

Finally, the authors review studies on the electrical properties of hybrid polymers which revealed that their electrical conductivity increased markedly with the proportion of pyrrole in the initial mixture. This is attributed to the formation of an electrically conducting polymer network in the non-conducting methacrylate matrix.

Chapter 1 - Electrically conductive textiles command a coveted place in the technical textile sector due to their suitability for some of the niche areas. They play an important role in the following areas: Static protection and electrostatic discharge (ESD) protection, electromagnetic interference (EMI) shielding, microwave attenuation, resistive heat generating textiles and interactive/electronic textiles. Conventional conductive textiles most commonly rely on metal based technologies wherein metallic materials are introduced in the form of fibre, filament and coatings onto textile substrates. Although the conductivity of the resultant textile substrate can be modulated from low to extremely high range in relation to the target application, there are certain limitations of metal based conductive textiles. Oftentimes, the manufacturing processes are complex and costly. Further, the desired qualities such as handle, aesthetic appeal and comfort properties are compromised together with weight burden and inflexibility. Conjugated polymers (CP) have emerged as one of the possible alternatives to metal based textiles in certain application areas. Textiles in all its forms have been explored for inclusion of CPs in order to combine the dual advantage of electric conductivity and flexibility of the resultant composite.

In this chapter basic overview of the polypyrrole synthesis by different methods such as chemical polymerization, Electro-polymerization, Photo initiated (UV) polymerization and γ irradiated polymerization and the science behind its formation is presented. The principle function of polypyrrole towards conductivity in textiles is discussed at length. The structure and morphology of polypyrrole prepared by all these methods has been investigated using the methods of FTIR, X - Ray diffraction and SEM. The conductivity was tested by two probe method. In order to assess long-term usage potential for practical applications, the polymerized fabrics were analyzed for their stability towards time and temperature. It was found that addition of sulphonic acid dopants had significant effect on conductivity and atmospheric aging of polypyrrole coated fabrics. Use of Intrinsically conductive polymers incorporated textile substrates was explored in the field of interactive/Smart textiles.

Chapter 2 - Among the heterocyclic systems, pyrrole skeleton is widely distributed in many natural and biologically relevant molecules such as alkaloids, porphyrins, and co-enzymes. Pyrrole derivatives play a crucial role in organic chemistry, medicinal chemistry, and heterocyclic chemistry. Pyrrole scaffold is extensively used in the synthesis of drug molecules with various pharmacological properties as well as in material sciences. Several synthetic protocols have been reported by enthusiastic researchers worldwide during the last decade.

Preface

In view of several environmental and regulatory issues, promoting sustainable green synthetic processes has become a global necessity. This paradigm shift from classical organic reactions to eco-friendly reactions as well as technologies has led to the evolution of green chemistry. However, for the development of environmentally benign reactions, the selection of alternate nontoxic solvent medium plays a significant role.

Water, a universal solvent, which is an essential element of life on our planet, is involved in different biological processes in nature. Despite its potential role in nature, due to the complexity of present-day organic reactions and the poor solubility of organic compounds in water, its role as a sole solvent in conducting organic reactions is limited. However, several researchers, with an interest to develop environmentally acceptable reactions, have reported aqueous phase reactions. In view of immense attention shown globally towards water medium reactions, an attempt is made to review the scientific research reports pertaining to the "Recent Research Advances in the Aqueous Phase Synthesis of Pyrroles" during the last decade.

Chapter 3 - Electrically-conducting polymers such as polypyrrole have been the focus of intense research interest over the last ten decades because they can be used in a wide range of technologies. However, it is well established that electrically-conducting polymers fabricated by chemical or electrochemical polymerization processes are either powders or intractable polymers which exhibit deficient mechanical properties and are difficult to process. A possible way of improving the mechanical properties of polypyrrole is by mixing it with other polymers in order to reach a synergetic overall performance. In this study pyrrole was blended with a methacrylate resin and the mixtures were processed by photopolymerization. This polymer processing method has the advantage that permits the incorporation of different additives and flexibilizers into the resins thereby optimizing its manufacture and improving the mechanical properties of the final cured material. The photoinitiator systems used to cure the mixtures pyrrole/methacrylate consisted of the iodonium salt p-(Octyloxyphenyl)phenyliodonium hexafluoroantimonate (IODS), in combination with Benzil α,α-dimethyl acetal (BDMA), α-

Methoxy-α-phenylacetophenone (MPAP) or the pair camphorquinone (CQ)/ethyl-4-dimethylamino benzoate (EDMB). Mixtures photoactivated with the IODS salt in combination with BDMA or MPAP were efficiently cured under UV irradiation (λ=365 nm). On the other hand, in mixtures photoactivated with IODS/CQ/EDMB and irradiated with visible light ((λ=470 nm) the polymerization of both methacrylate and pyrrole was much slower. Scanning electron microscopy studies showed no sign of phase separation demonstrating that the pyrrole/methacrylate blends formed an interpenetrating network. Studies of electrical properties of the hybrid polymers revealed that their electrical conductivity increased markedly with the proportion of pyrrole in the initial mixture. This is attributed to the formation of an electrically conducting polymer network in the non-conducting methacrylate matrix.

In: Pyrrole: Synthesis and Applications
Editor: Colin Welch
ISBN: 978-1-53617-137-2
© 2020 Nova Science Publishers, Inc.

Chapter 1

POLYPYRROLE BASED FABRICS: SYNTHESIS AND APPLICATIONS (REVIEW)

Smita Deogaonkar-Baride[*], *Narendra V. Bhat, Padma Vankar and Anjan K. Mukhopadhyay*
The Bombay Textile Research Association,
Mumbai, Maharashtra, India

ABSTRACT

Electrically conductive textiles command a coveted place in the technical textile sector due to their suitability for some of the niche areas. They play an important role in the following areas: Static protection and electrostatic discharge (ESD) protection, electromagnetic interference (EMI) shielding, microwave attenuation, resistive heat generating textiles and interactive/electronic textiles. Conventional conductive textiles most commonly rely on metal based technologies wherein metallic materials are introduced in the form of fibre, filament and coatings onto textile substrates. Although the conductivity of the resultant textile substrate can be modulated from low to extremely high range in relation to the target application, there are certain limitations of metal based conductive

---

[*] Corresponding Author's Email: conductive@btraindia.com.

textiles. Oftentimes, the manufacturing processes are complex and costly. Further, the desired qualities such as handle, aesthetic appeal and comfort properties are compromised together with weight burden and inflexibility. Conjugated polymers (CP) have emerged as one of the possible alternatives to metal based textiles in certain application areas. Textiles in all its forms have been explored for inclusion of CPs in order to combine the dual advantage of electric conductivity and flexibility of the resultant composite.

In this chapter basic overview of the polypyrrole synthesis by different methods such as chemical polymerization, Electro-polymerization, Photo initiated (UV) polymerization and γ irradiated polymerization and the science behind its formation is presented. The principle function of polypyrrole towards conductivity in textiles is discussed at length. The structure and morphology of polypyrrole prepared by all these methods has been investigated using the methods of FTIR, X - Ray diffraction and SEM. The conductivity was tested by two probe method. In order to assess long-term usage potential for practical applications, the polymerized fabrics were analyzed for their stability towards time and temperature. It was found that addition of sulphonic acid dopants had significant effect on conductivity and atmospheric aging of polypyrrole coated fabrics. Use of Intrinsically conductive polymers incorporated textile substrates was explored in the field of interactive/Smart textiles.

Keywords: conductive textiles, polypyrrole, conductivity, atmospheric ageing, interactive textiles

1. INTRODUCTION

The enormous development in the field of material science in terms of smart and functional materials and their merging into textiles has resulted into textile materials with smart functions, high performance and novel functionalities. Imparting electrical conductivity to textiles is one of the functionalities that assume significant importance in application areas of static and electrostatic discharge (ESD) protection [1], electromagnetic interference (EMI) shielding [2], heat generating textiles [3], and microwave attenuation [4]. These applications of conductive textiles cater to diverse areas ranging from mundane domestic items such as static shock–proof carpets to static dissipative garments and accessories for electronics & semiconductor industry and high tech applications such as stealth technology for defence establishments. The method of manufacturing of traditional conductive textiles rely on metal based techniques, which adds weight burden, offer metallic properties detrimental to feel, comfort and aesthetic aspects. These shortcomings of metal based textiles can be overcome by incorporating textiles with a intrinsically conductive polymers (ICP's) which possess a wide range of conductivity, easy processing, biocompatibility and environmental stability. This class of polymers is inherently electrically conductive due to the presence of alternate conjugated single and double bonds in their structure. The peculiar bonds arrangement however, restricts the flexibility of intermolecular chains increasing the structural rigidity. Hence, the CPs are usually combined with a foreign substrate to form a composite structure. Textiles in all its forms have been explored for inclusion of CPs in order to combine the dual advantage of electric conductivity and flexibility of the resultant composites [2, 5, 6].

Among the ICPs, Polypyrrole has received more attention as subject of research due to ease in synthesis, highly conducting nature and low cost of monomer [7, 8].

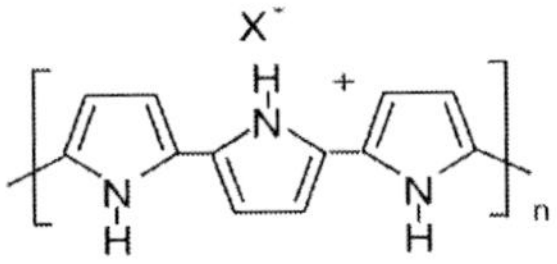

Figure 1. Chemical structure of polypyrrole.

Polypyrrole (PPy) is a chain of pyrrole units - a heterocyclic ring with nitrogen atom (Figure 1). The polymeric structure consists of pyrrole unit joined to successive units by alternating to 180° to give α-α bonding. The 2, 5- coupling (i.e., α-α coupling) ensure reversal of unit's thereby maintaining planarity and linearity of chains. Any linkages apart from this bonding create deficiencies in polymer chain structure leading to structural defect and the corresponding deterioration of conductive properties by breaking the planarity. The hetero aromatic and extended p-conjugated backbone structure of PPy provide it with chemical stability and electrical conductivity, respectively.

According to Diaz and co-workers, during polymerization of Pyrrole, the initiation takes place with oxidation of a monomer unit leading to formation of a radical cation. Two such radical cations combine and undergo deprotonation to form a stable aromatic dimer. The propagation step involves further oxidation of dimer yielding a reactive species. It combines with the monomer cation available abundant in the vicinity and thus oligomer is formed. The continuous oxidation, combination and deprotonation results in the growth of oligomer chains and finally leads to precipitation of polymer in the solution (Figure 2 F). The propagation step continues until the radical becomes too weak or the growth is hindered by steric reasons. As mentioned earlier, one interesting observation is that the each pyrrole unit joins with the successive unit by alternating to give alpha-alpha bonding. Further, according to the reported literature, the mechanism of chemical and electrochemical polymerization has similar steps [9].

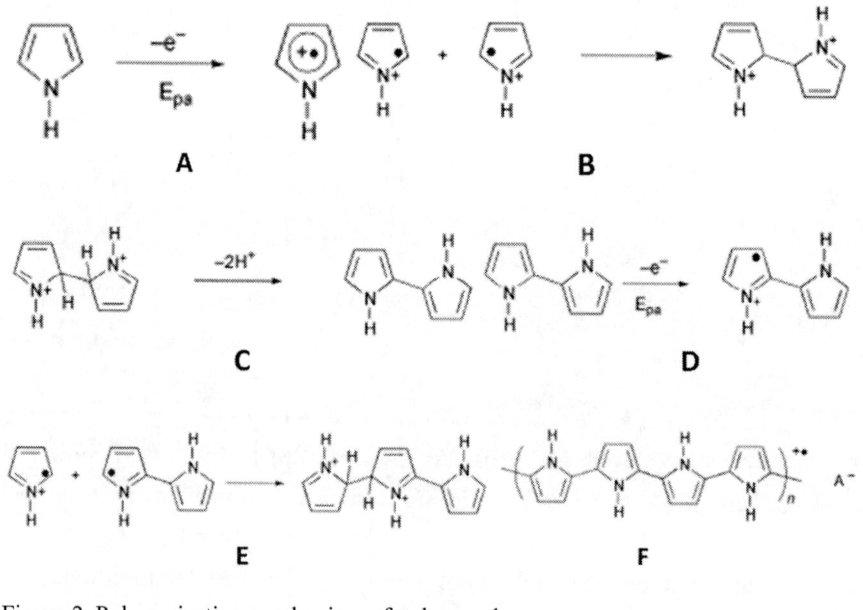

Figure 2. Polymerization mechanism of polypyrrole.

1.1. Chemical Polymerization of Pyrrole

Chemical synthesis of polypyrrole proceeds via the oxidation of pyrrole with an oxidants such as ferric chloride [10], ammonium peroxidisulphate [11] etc. The oxidation potential of Pyrrole is 0.8V, which is lower than other heterocyclic monomers. Consequently it is readily synthesized from range of aqueous and non-aqueous solvents. During chemical polymerization of pyrrole, electro neutrality of the polymer matrix is maintained by incorporation of anions from the reaction solution. These counter ions are usually anions of the chemical oxidant or reduced product of oxidant. Many researchers reported chemical polymerization of pyrrole using different oxidants such as Ferric chloride [7, 10], Ammonium peroxidisulphate [2, 11, 12], Hydrogen peroxide [13] and halogens [14]. Chao and March reported that pyrrole oxidative polymerization reaction can be carried out using a variety of oxidative transition metal ions (Fe^{3+}, Cu^{2+}, Cr^{6+}, Ce^{4+}, Ru^{3+} and Mn^{7+}) which forms a conductive cationic radical as end product [15]. The use of mixtures of two different oxidants has also proven useful to get improved conductivity. Polymerization of pyrrole in presence of surfactants such as naphthalene sulphonic acid sodium salt (NSA), dodecyl benzene sulphonic acid sodium salt (DBSA), anthraquinone sulphonic acid sodium dalt (AQSA) leads to increase the mass yield due to incorporation of salt in to the polymer [10, 16]. Cationic surfactants were found to inhibit the polymerization of polypyrrole [17]. In situ chemical polymerization of polypyrrole on variable substrates by a process similar to textile dyeing/finishing was already discussed in our previous work [10, 11, 18]. Here the reaction was carried out at a temperature of 4-5°C, to reduce the number of side reactions. Polypyrrole conductivity as a function of temperature is well discussed in our earlier studies [10].

1.2. Electrochemical Polymerization

Electrochemical oxidation of pyrrole forms a film of polypyrrole at the electrode surface [19]. It proceeds in three following steps: Initiation (involves monomer radical cation formation), Propagation (Combination of radical cation followed by loss of hydrogen ions to continue the chain propagation in polypyrrole) and Termination (occurs when no monomer is present for oxidative polymerization or side reactions terminate the polymer chain) [20]. Potentiostatic (constant potential), Cyclic voltametry [21] and galvanostatic (constant current) methods can be used to polymerize the pyrrole electrochemically. In our study, we followed Potentiostatic method to coat polypyrrole film over chemically synthesized PPy coated fabric. It gives uniform coating of polypyrrole with improved conductivity.

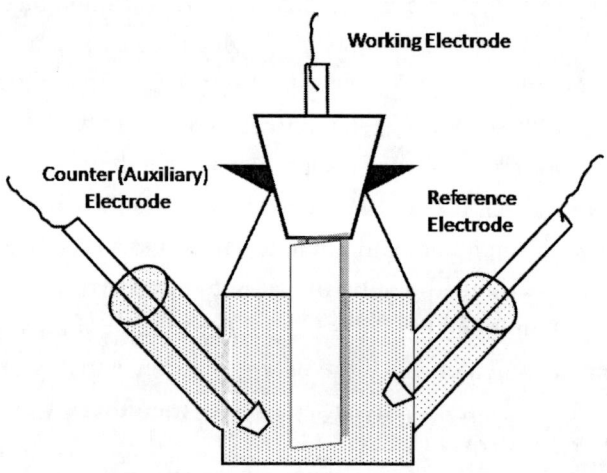

Figure 3. Electro polymerization cell.

1.3. Gamma Irradiation Polymerization

Gamma irradiation polymerization has been used widely to generate novel polymeric materials that have newer properties. This method helps to prepare materials at room temperature and at normal pressures. Gamma irradiation can be easily controlled and nowadays different sources are available. In addition as it does not need any initiator and therefore the process does not introduce any impurities into the materials [22–25]. The main advantage of this route for the formation of PPy, is that it can be obtained in the pure form. The results of this study will reveal that highly uniform and nanometer sized globules can be obtained. This as well as electron irradiation processes give materials with high adherence to the substrate.

1.4. Photo Initiated (UV) Polymerization

Photochemical reactions are based on the absorption of light that excites the electrons of a molecule and can lead to dissociation, isomerization, abstraction, electron or energy transfer, and bond formation. These reactions have been the subject of many studies in various fields including organic chemistry, molecular biology, electronics etc. Photo induced chemical reactions can advantageously be used in the field of polymer chemistry. Among them, photoinitiated polymerization has attracted tremendous technological interest in the past few decades because of its wide use in many applications including surface coatings, printing inks, adhesives, microelectronics, printing plates, three- dimensional imaging and microfabrication processes [26]. In the free radical mechanism, polymerization can be initiated by both cleavage (type I) and hydrogen abstraction type (type II) initiators [27-29]. The radical generation processes of type I and type II photo initiators are entirely different. While type I photo initiators generate initiating radicals by a unimolecular cleavage reaction, type II initiators undergo bimolecular hydrogen abstraction reactions (Figure 4).

UV lamps capable of producing high power radiation with availability of wavelengths from the near UV (354 nm) to the deep UV (126 nm) have been developed [30, 31]. The development of such novel lamps gives enormous potential for materials processing. Several applications by using

UV have already been established and include material deposition, polymer etching and surface modification [32]. The experimental results have shown that excimer VUV and UV sources are ideal for initiating the photo-polymerization processes because of very rapid polymerization at ambient temperature.

Photo initiated (UV) polymerization is attractive in photolithographic application of this polymer, since it allows alterations in synthesized PPy morphology by change of excitation light wavelength [33] and theoretically it might be applied for the design of electronic chips. However, because of slow rates this type of polymerization has not been used much as for the preparation of conducting polymers as compared with chemical or electrochemical polymerization.

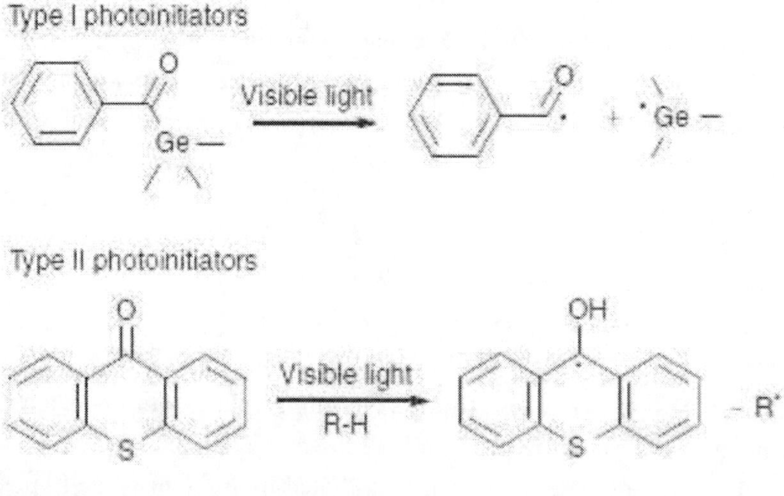

Figure 4. Photochemical Polymerization.

The polymers produced by all these methods are generally in powder form or solid sheet. These forms cannot be used directly for making practical applications. Hence to make practical utilisation of conductive polymers, researchers have been looking at a number of different methods. Their processability is reviewed below to highlight their potential uses in different fields. It includes direct polymerization on to polymers sheets, metal sheets, glasslates, clays, porous materials, textiles fibers/fabrics etc

[34, 35]. In one strategy the oxidant (ferric chloride or ammonium peroxidisulphate) is mixed with the substrate (such as polyvinyl acetate, polyacrylate, polyethylene oxide or rubber) and then exposed to pyrrole vapours [36-37]. In some cases, the method varies by soaking the substrate in pyrrole solution and then immersing it in an oxidant solution [38-39]. Surfaces of substrates have been functionalized with plasma treatment/alkali treatment to enhance polypyrrole-substrate adhesion (interfacial bonding) and conductivity. Surface treatment by atmospheric pressure plasma glow discharge (APGD) was performed by Garg et al. to improve abrasion resistance and conductivity of PPy coated Polyester and wool fabrics [40]. Polypyrrole can be directly deposited on to the substrate surface by placing the object in solution containing pyrrole and oxidant. Composites have been made in which organic materials and inorganic oxides or salts of different metals viz SnO_2 [41], TiO_2 [42], Y_2O_3 [43], Fe_3O_4 [44] etc combine in special manner with the Polypyrrole. In some cases it have been made by dispersing PPy powder in melted LDPE, HDPE or by dispersing it in silicone rubber or vinyl ester and curing the material [45-46]. Conductive Fabric composites such as PPy-PET and PPY-Jute composites were prepared by Khare et al. for EMI shielding application, to overcome atmospheric ageing problems related to polypyrrole based fabrics [35]. Graft co-polymerization of pyrrole was also followed by many researches to form soluble or process able polypyrrole [47]. Plasma polymerization of pyrrole at low pressure has already been reported, where the injection of a pyrrole aerosol into variable plasma and in situ doping by co injection of NOBF4 or iodine, results in polymerization and coating deposition of polypyrrole [48].

The present review chapter describes the synthesis of polypyrrole by all the four methods discussed above. The structure and morphology of polypyrrole prepared by all these methods has been investigated using the methods of FTIR, X - Ray diffraction and SEM.

2. Experimental

2.1. Preparation of PPy Coated Fabrics by Chemical Polymerization

In-situ chemical polymerization of PPy on fabric sample was carried out as two step process in a laboratory grade jigger machine. The detailed description of the polymerization set up has been given in our previous work [11]. The fabric samples were scoured and cleaned prior to polymerization. The reaction was carried out at a temperature of 4-5°C. Variable fabrics (cotton, Polyester, Polyester-Cotton blend) were treated with pyrrole and ferric chloride hexahydrate subsequently to get PPy coated Cotton, Polyester and Polyester –Cotton blend fabric respectively. Concentration of monomer solution of pyrrole was varied at 10, 20 and 30% on weight of the fabric. The treatment time in the monomer solution was kept as 1 hr. for proper adsorption. Further, the pre–cooled oxidant, ferric chloride hexahydrate was added to same bath as a second step. The monomer to oxidant ratio of 1:2.33 was maintained for the entire set of experiments. The final material to liquor ratio was maintained at 1: 40. After completion of the treatment fabric sample was removed, washed with distilled water and dried over night. Polypyrrole coated fabrics thus obtained were characterized for various properties.

2.1.1. Plasma Treatment Followed by Chemical Polymerization

Atmospheric pressure plasma treatment (generated through dielectric-barrier discharge) has been applied to synthetic and natural fabrics to improve surface adhesion and increase the amount of PPy deposition over the substrate, thereby increasing conductivity and durability. We used such pre-treatments with plasma for the cotton fabrics, before they were given the usual chemical treatment with pyrrole. On a few occasions polymerization of pyrrole was carried out simultaneously in the plasma chamber and polypyrrole was deposited/ grafted directly on the cotton fabrics. The deposition was thin but enough to change the properties. Among various gases He+O, He+Ar+O and He plasma were applied here to improve the adhesion and rate of polymerization. Properties such as surface resistivity, abrasion resistance were measured for the coated fabrics.

2.2. Preparation of PPy Coated Fabrics by Electro-Chemical Polymerization

For electrochemical synthesis "Gamry potentistat/galvalostat"-reference 3000 model was employed. The chemically polymerized cotton fabrics were utilized as working electrode onto which the deposition of electrochemically polymerized PPy takes place. A platinum rod was used as a counter electrode. All the potential measurements were referred to Ag/AgCl reference electrode. The Potentiostatic synthesis was adopted at a voltage of 1.2 V. The electrochemical synthesis on top of the chemically polymerized fabric (3 X 8 cm) was carried out in aqueous sulfonic acid salt solutions containing pyrrole monomer at varying concentrations. All experiments were carried out at room temperature and without stirring.

Electrochemically produced coatings have certain merits such as superior conductivity, thermal stability and ageing resistance. Additionally, electrochemical polymerization offers opportunity to introduce functional molecules and doping agents during electrochemical polymerization offering composite substrate with altogether distinct properties.

2.3. Preparationof PPy by Gamma Irradiation Polymerization

Conducting polypyrrole (Ppy) can be synthesized by γ – radiation from Co^{60} source Fabric pieces of size 4 x 4 cm were pre soaked in dilute pyrrole solution and inserted in the chamber. Occasionally glass plates, Copper plates and fabric pieces of same sizes were kept in a petri dish containing pyrrole solution. The important advantage of this method is oxidant materials like $FeCl_3$ can be avoided and therefore the Ppy obtained is in highly pure form. Polypyrrole particles obtained had a size much less than that obtained by chemical and electrochemical polymerization techniques [49].

Gamma-ray irradiation offers several advantages for the initiation of polymerization with organic compounds over the conventional chemical methods. γ-radiation has been applied extensively for the initiation of polymerization, grafting of polymer chains onto polymeric backbones, modification of polymer blends and preparation of interpenetrating polymer networks [50]. A large number of hydrated organic radicals including aq, OH, H_3O^+ and pyrrolium cation were produced during γ-ray irradiation in aqueous solution. These hydrated radicals initiated the polymerization of free pyrrole existing in the reaction solution.

2.4. Preparation of PPy by Photo Initiated (UV) Polymerization

For UV-photo processing, a precursor was prepared by using pyrrole and 1 M H_2O_2 in the ratio of 1:1. The solutions were mixed together by stirring for 5 min. The mixture was then poured in optically flat petry dishes in which pieces of Glass plate, Cotton fabrics and Copper foil of size 4x4 cm each were kept dipped in the solution. This allowed the polymerization directly on the substrate to be studied. The solution was then exposed to UV radiation from a UV lamp (wavelength of 200 nm) for 6 hrs. The distance between the sample and lamp was 10 cm, with the lamp having 100 W of power. It was noticed that the sample surface temperature did not rise much. The thin film of polypyrrole started depositing on various substrates after nearly 2 hours and sufficiently thick film was formed after 6 hours. The substrates were then taken out and washed thoroughly by distilled water several times. The fabrics and films were dried for 1 day before testing.

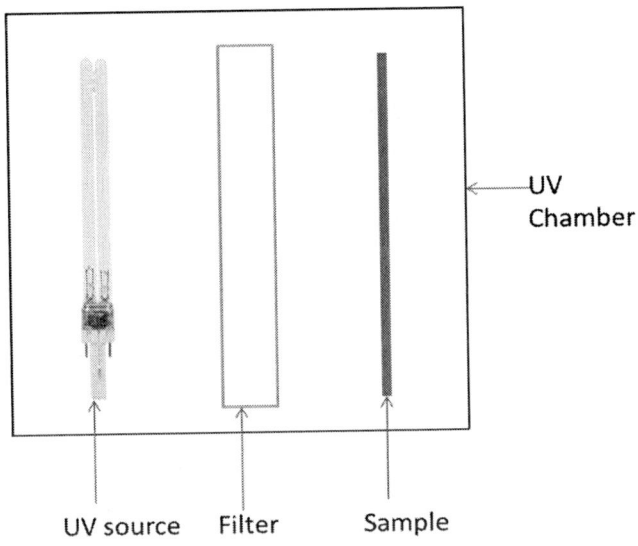

Figure 5. UV chamber for photo processing of pyrrole.

3. CHARACTERIZATION

The electrical conductivity of PPy coated fabrics were measured in terms of electrical surface resistivity by employing AATCC-76:2005 standard method. The unit of surface resistivity is ohm/square (ohm/sqr.) The morphology of PPy deposition was observed by scanning electron micrographs (SEM) on JEOL-JSM-5400 model. The tensile strength (in warp direction) was carried out as per ASTM D 5035:06. FTIR is a technique employed for chemical characterization of a substrate. The ATR-FTIR spectra were recorded at a resolution of 4cm^{-1} using a Perkin-Elmer spectrometer, model system FTIR-Spectrum Two. Attenuated (ATR) spectra were obtained using ATR MIRacle Diamond crystal, horizontal accessory. The scanning range used was 4000-650cm^{-1} and an average scan of 8 scans was recorded. X-ray diffraction analysis was carried out using X-ray diffractometer of model X-pert Pro. Scan was obtained in the 2θ range of 5^0 to 60^0 using CuKα radiation source.

4. RESULTS AND DISCUSSION

4.1. Chemical Polymerization

4.1.1. Electrical Characteristics of PPy Coated Fabrics

In-situ polymerization of polypyrrole on variable substrates (Cotton, Polyester and Polyester/Cotton blend) in dynamic conditions consisted of the initial monomer treatment followed by controlled addition of oxidant. Conductivity of the fabrics achieved corresponds to the presence of Cl^- dopant ions. These chemically synthesized PPy coated fabric shows the surface resistivity in the range of 20–1000 ohms/square for cotton [10, 11], 300-5000 ohms/square for polyester [52] and 50-2500 ohms/square for Polyester/cotton blend fabric. Due to its tuneable conducting nature it can be used in wide range of applications such as, low conductive fabrics for antistatic clothing, medium conducting fabric for smart textiles and highly conductive fabric for EMI Shielding. Also it was observed that, PPy coated cotton fabric posses more conductivity than PPy coated Polyester and polyester-cotton blend fabric. This is due to higher surface energy of cotton fabric than polyester-cotton blend and polyester fabric. In case of cotton, both absorption and adsorption of polypyrrole takes place which leads to more conductivity than polyester fabric where only adsorption of PPy takes place.

In case of plasma treatment followed by chemical polymerization, it was found that He+O, plasma gas lowered the resistivity value of PPy coated polyester fabric compared to the plasma untreated-PPy coated PET samples (Figure 6a). The surface resistivity of plasma untreated PPy coated polyester sample was 1400ohms/square (at 20% Py concentration o.w.f) and same for plasma pre treated sample is 498ohm/square. After abrasion up to 2000 cycles, plasma treated PPy coated sample shows the change in surface resistivity up to 700 ohms/square (i.e., 40% loss) which is less compared to change from 1400 to 2400 ohms/square in case of plasma untreated PPY coated samples (i.e.,72% Loss). Tonal change results also clearly shows that the plasma treatment was very effective at improving the adhesion strength between the PET substrate and conductive PPy coating (Figure 6b). Thus improved adhesion has made the plasma treated PET more conductive.

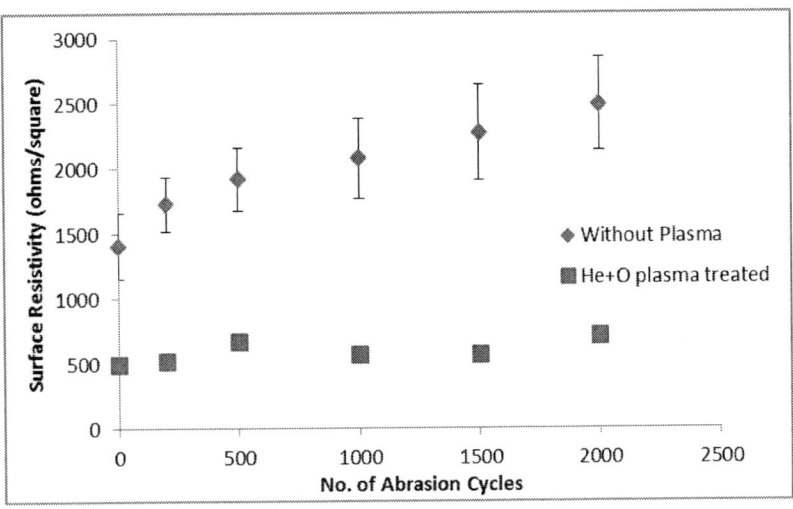

Figure 6a. Surface resistivity results of PPy-coated PET fabric with and without plasma treatment.

Source: Deogaonkar S. C, Improvement of Polypyrrole coating adhesion on polyester fabric by atmospheric pressure plasma technology, colourage June 2018, Permission asked.

Figure 6b. PET Samples after 2000cycles of abrasion (a) Control (b) He + O.

4.1.2. Morphological Study

The SEM images of untreated cotton fabric and polypyrrole coated cotton fabrics are shown in Figure 7. The untreated fabric shows a clear surface with cylindrical fibres and PPy coated fabric shows granular deposition of polymer. The polymerization of polypyrrole on cotton fabrics takes place through diffusion of polymer inside the fibre bulk as well as deposition on the fibre surface and the interstices in the fabric.

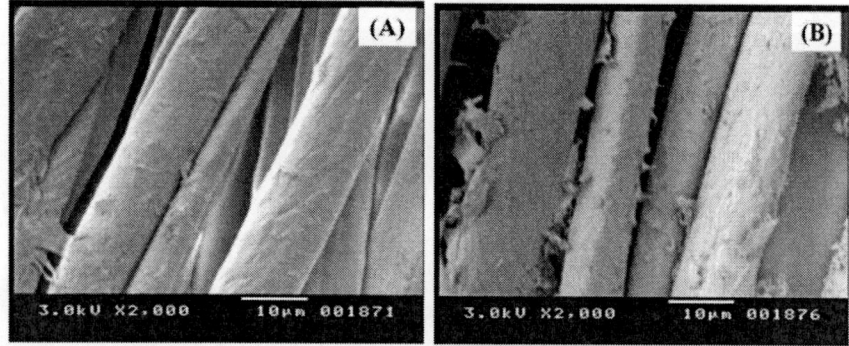

Patil A. J and Pandey A, Indian J Fib Tex. R, 37, 2012, 107-113 with permission.

Figure 7. SEM micrographs of PPy coated cotton fabrics prepared with chemical synthesis.

4.1.3. FTIR Spectroscopy

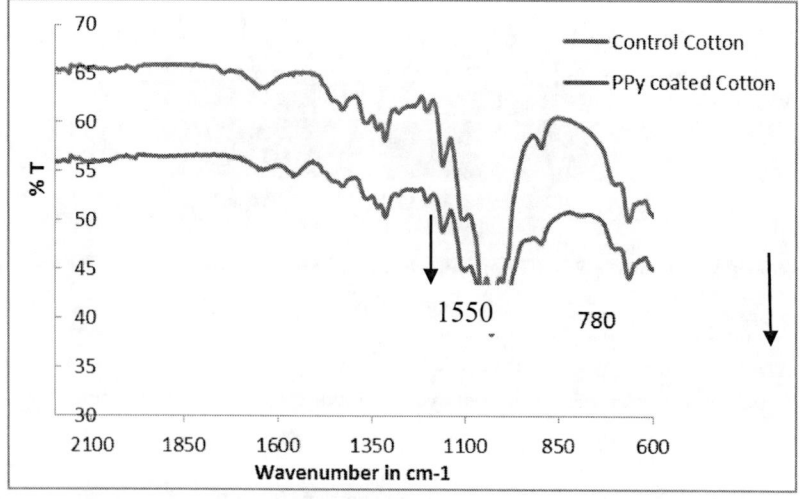

Figure 8. FTIR-ATR spectra of PPy coated cotton fabric.

FTIR-ATR spectra of untreated cotton fabric and chemically synthesized PPy coated cotton fabrics shown in Figure 8. Control cotton shows characteristic bands at 3323 (OH stretching), 2953 (CH_2 antisymmetric stretching), 2922 (C-H stretching of cellulose), 1740 (C-O linkage of lactone ring), 1470 (OH in plane bonding), 1338 (OH in plane bending), 1169 & 1113 (C-O-C, asymmetric bridge), 1061 (asymmetric in-plane ring stretch). The prominent bands of polypyrrole are observed in FTIR spectra of treated fabric. For instance, aromatic ring vibrations can be seen 1550 cm^{-1}. Similarly C-N in plane deformation at 1300 cm^{-1} and C-H in plane vibrations at 1170, is also noted. This proves the presence of PPy polymer onto chemically synthesized cotton fabrics.

4.1.4. XRD Studies (Crystallinity and Molecular Order)

Several studies have investigated the manner in which polymer chains are arranged in the solid state for polypyrrole, with using the tool of x-ray diffraction. In general XRD provides information on the crystalline and amorphous structure of the polymers, its spatial order and orientation of chains. It has been reported that most of the PPy synthesized by Chemical and electrochemical methods, doped by inorganic counter ions (Cl^-, ClO_4^-, TS^- etc.) are essentially amorphous [56]. The Mitchell and Geri revealed a molecular anisotropy due to all trans coupled PPy chains lying in planes parallel to the electrode surface, but randomly oriented in the direction perpendicular to the electrode [66]. Also it was revealed that, anisotropy was observed when planar dopants were used. The increase in degree of anisotropy was found to favouring an increase in electrical conductivity of PPy. Polymerization conditions that favour high conductivity also produce high degree of Crystallinity (up to 37%) and larger crystallite domain length (up to 2.6nm or 8 pyrrole rings) [9]. Davidson and co-workers found clear evidence of a helical structure produced from all cis coupling of pyrrole ring [67].

Here, the X-ray diifractogram of PPy coated cotton fabric shows two weak peaks at 15 and 16.5 degrees and a very strong peak at 22 degrees (Figure 9a). These are due to cotton fabrics. In the region of 26–27 degrees

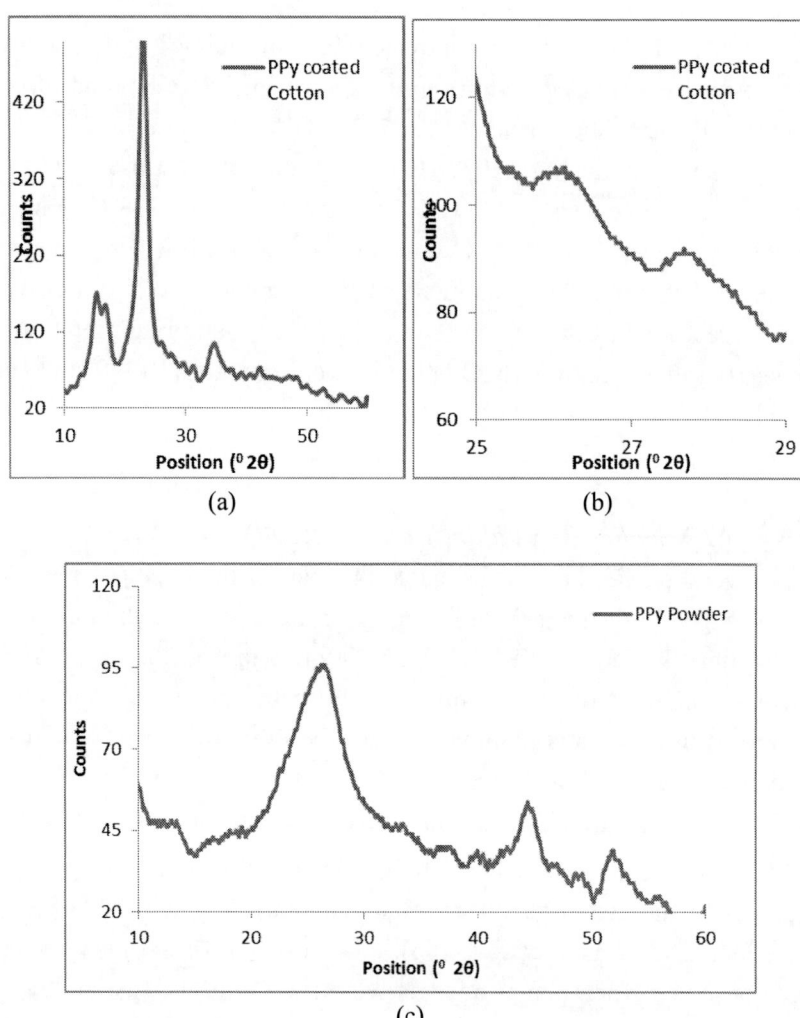

Figure 9. XRD of Chemically synthesized a) PPy coated cotton fabric (Ovarall spectrum), b) PPy coated cotton fabric (magnified view) and of c) Pure PPy powder.

very weak peaks/ humps are observed which are due to polypyrrole. The intensity of these humps is very weak in comparison to cotton because cotton is highly crystalline material whereas polypyrrole is amorphous-like material. In addition much of polypyrrole is embedded in the cotton as cotton is porous and absorbs pyrrole monomer before polymerization. These humps are very clear in the re-plot shown in the same figure as magnified view (Figure 9b). This is in conformity with the diffraction data reported earlier for the composite as well as polypyrrole powder alone [66, 69]. This shows that polypyrrole has been indeed deposited on the cotton fabric in or samples.

4.1.5. Mechanical Properties

The effect of in-situ polymerization of polypyrrole on tensile properties of cotton fabrics was studied in terms of warp way tensile strength. Figure 10 shows the behaviour of warp way tensile strength of PPy coated cotton fabrics at different concentrations of monomer and fixed duration of polymerization. The polymerization carried out at 10(%) concentration of pyrrole has not resulted in significant reduction of strength. However, at higher concentrations of 20(%) and 30(%) pyrrole, there is decrease in strength to the tune of 15-20(%). This loss can be attributed to the inflexible and rigid nature of polypyrrole coatings formed on the substrate which limits the contribution of fibre to fibre cohesion toward tensile stress. The oxidative degradation of cotton could also have played a part as ferric chloride is a strong oxidant.

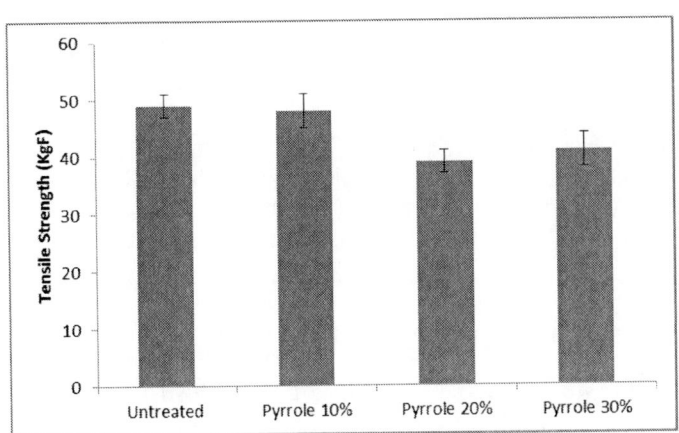

Source: Patil A. J. and Pandey A., Indian J Fib Tex. R, 37, 2012, 107-113 with Permission.

Figure 10. Tensile strength of PPy coated cotton fabrics prepared by varying the concentration of pyrrole Untreated; 10(%) pyrrole; 20(%) pyrrole; 30(%) pyrrole.

4.2. Electrochemical Polymerization

4.2.1. Electrical Characteristics

In electrochemical polymerization, although the concentration of pyrrole was varied the monomer to dopant ratio was kept constant at 1:0.4. From Figure 11, it can be seen that the electrochemical polymerization was not initiated at dilute concentrations of monomer (i.e., 0.025M and 0.038M). However, the anode force seems to be sufficient for the movement of monomer molecules at 0.05M concentration. This has resulted in remarkable decrease in the surface resistivity due to build up molecular chains and subsequent deposition of polymer formed. Further, with increasing concentration beyond 0.05M has not yielded any appreciable decrease in the surface resistivity.

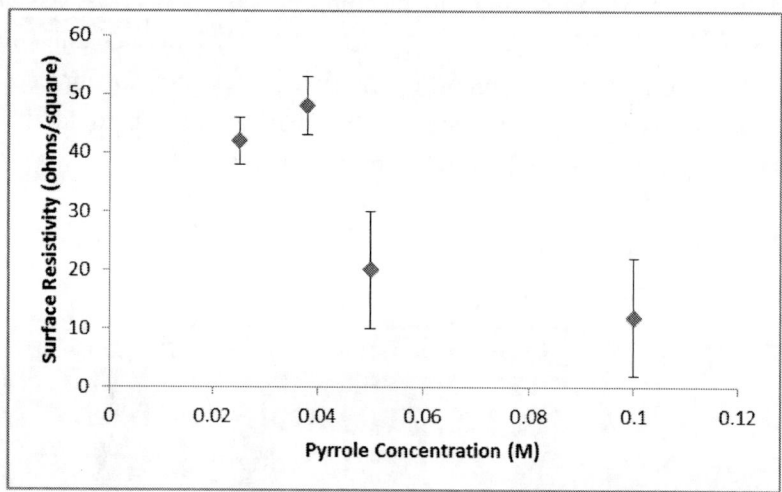

Figure 11. Effect of monomer concentration on electrical surface resistivity.

In order to analyze the effect of dopant structure on the resultant electrical conductivity of fabrics during electrochemical polymerization, a monomer concentration of 0.1M was used with 1:0.4 monomers to dopant ratio. Figure 12 shows the percent change in the surface resistivity for different dopants. It is worth noting that the variation in dopant structure seems to have no significant impact on the conductivity of the resultant fabrics.

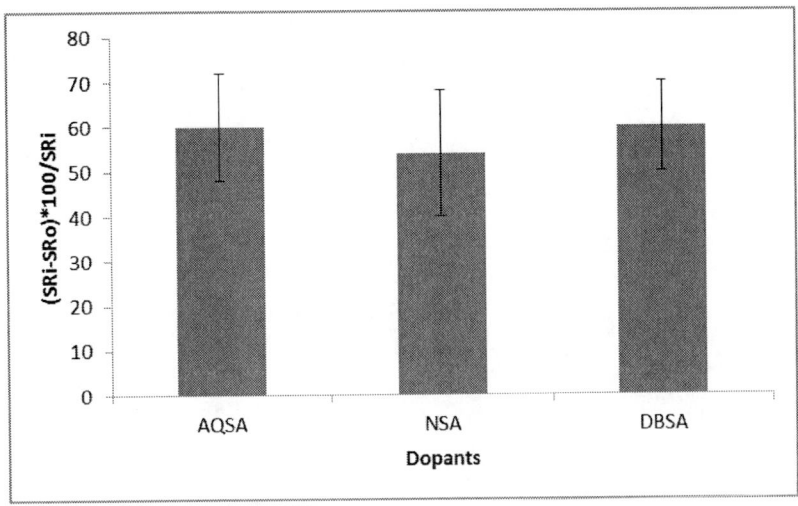

Figure 12. Effect of dopant structure on electrical surface resistivity.

4.2.2. Morphological Studies

The morphology of polymer deposition in case of electrochemical polymerization was observed by SEM analysis. Figure shows the SEM images of PPy coatings onto fabric surfaces for each dopant synthesis. Figure 13A shows the surface morphology of fabrics before electrochemical polymerization. It is characterized by the presence of micro serrations on the fibre surface. Figure 13 B-D exhibits the surface morphology obtained by insertion of AQSA, NSA and DBSA dopant molecules respectively. It is worth noting from images that the AQSA dopant yields smoother surface in comparison with other dopants. Similar observation was reported by Kunh et al. [53]. The planer structure of AQSA is believed to be responsible for the smoother morphology. In case of other dopants, the polymer formation in granular form was observed.

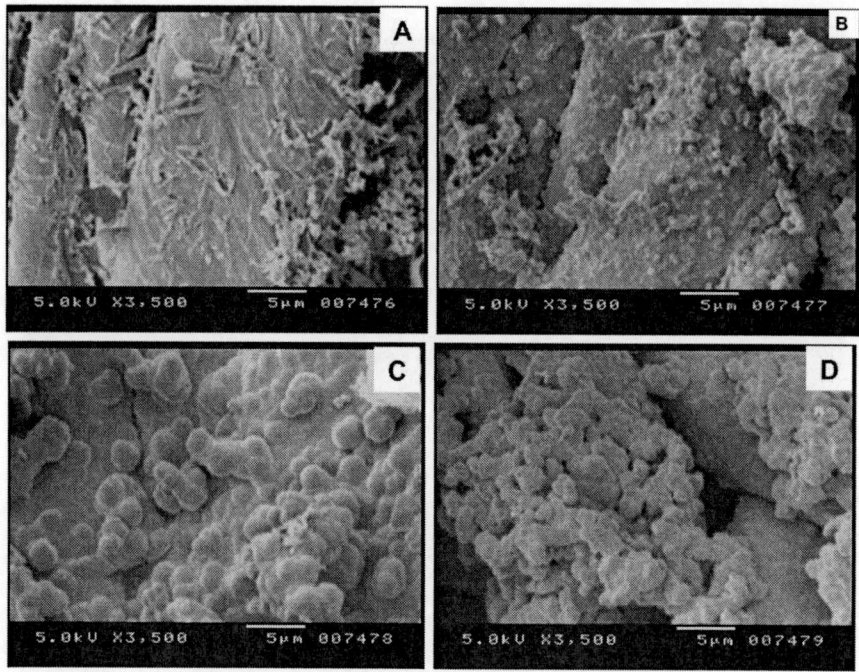

Figure 13. SEM images of fabric prepared by electrochemical polymerization. (A: control; B- AQSA dopant; C-NSA dopant; D- DBSA dopant).

4.2.3. FTIR Spectroscopy

Figure 14 depicts the ATR spectra of PPy coated cotton fabrics prepared by electrochemical polymerization with different dopants. The base cotton fabrics spectra shows the characteristic bands of PPy confirming the PPy polymeric coatings obtained by chemical polymerization. The bands at 1546 and 1458 correspond to pyrrole ring vibrations. The C-N stretch vibrations can be noticed at 1285 cm^{-1}. It is worth noting that the presence of all bands corresponding to PPy can be observed after electrochemical polymerization. This can be ascribed to the consolidation of polymeric coatings in the electrochemical synthesis. Further, the appearance of a sharp band at 1700 cm^{-1} may be attributed to the over oxidation of PPy during polymerization.

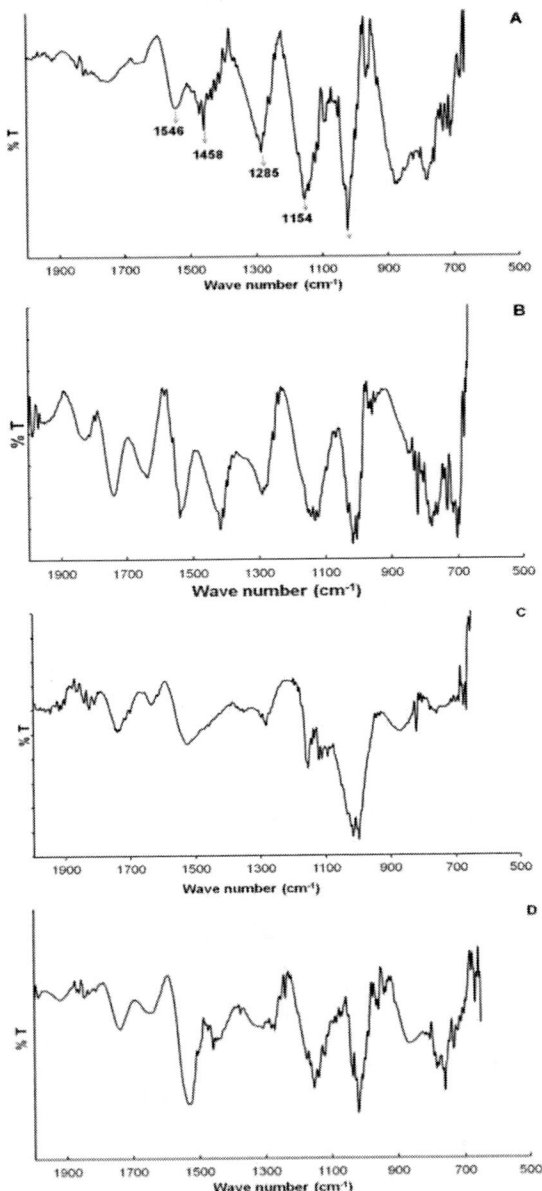

Related data presented by Patil A. J. and Deogaonkar S. C. at 53rd Joint technological Conference held at Mumbai, 2012.

Figure 14. FTIR-ATR spectra of electrochemical polymerized fabrics: A, Before electrochemical polymerization; B, 0.1M PPy/AQSA dopant; C, 0.1M PPy/DBSA dopant; D, 0.1M PPy/NSA dopant.

4.2.4. Polypyrrole Stability

The atmospheric oxidation with respect to time, causes chemical degradation of PPy by disrupting the continuity in conjugation length and subsequently hinders the charge movement as well as charge hopping at defect sites. The reaction of oxygen on PPy backbone leads to formation of α, β-unsaturated carbonyl groups which are believed to act as an electron trap.

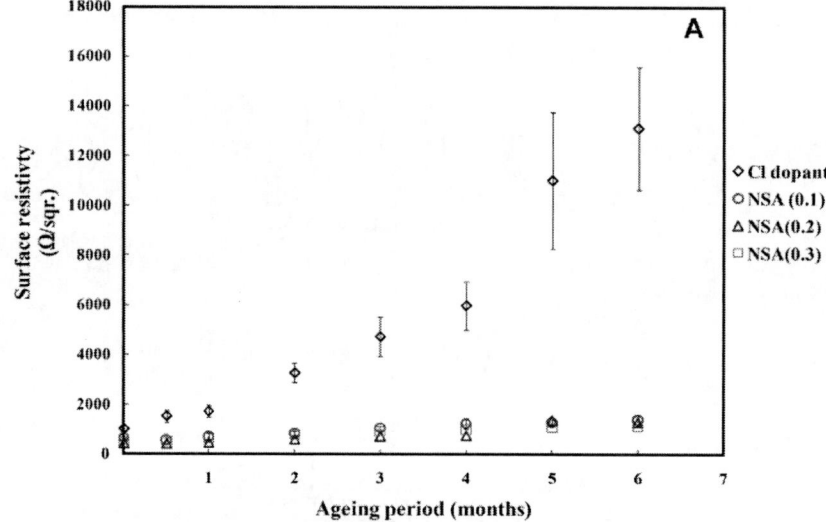

Source: Patil A. J. and Deogaonkar Smita C., 2012, J Appli Polym Sci., 125, 844, with Permission.

Figure 15. Ageing behaviour of PPy coated cotton fabrics: A) doped with Cl⁻ dopants and NSA dopants.

The addition of aromatic sulfonic acids makes significant difference in terms of improvement in conductivity and stability of PPy as reported in the literature [10]. The more planer structure of the sulfonic acid dopants was believed to be responsible for improvement in conductivity as it helps in inter-planer stacking of dopant molecules. This peculiar positioning of dopants between PPy chains facilitates interchain charge hopping more readily than small dopants such as Cl⁻ ions. Hence, the increased hopping opportunities contribute towards higher conductivity and improved stability towards atmospheric ageing.

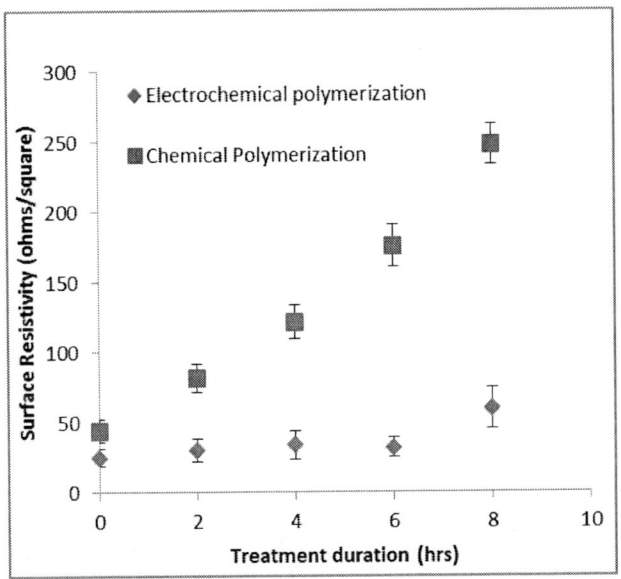

Figure 16. Effect of heat treatment on fabrics (Electrochemical polymerization conditions: 0.05M, AQSA- 0.02, duration of polymerization- 3h).

The heat stability of PPy coated fabrics before and after electrochemical polymerization was assessed by exposing fabrics in hot air oven at 100°C. The change in conductivity was monitored at an interval of 2 h with total exposure time of 8 h. It was observed that fabric with only chemically polymerized PPy performs worse than that of fabric coated with electrochemically polymerized PPy (Figure 16). The consolidation of polymer as well as presence of sulfonic acid dopants is believed to be responsible for superior heat resistance of electrochemically polymerized PPy-cotton fabrics.

The temperature and dopant ions have large role to play in the stability of polypyrrole [10, 54]. Solid state NMR of PPy films at 150^0C indicates the loss of dopant ion accompanies the loss of conductivity [55]. Dodecyl sulphate decomposes on heating possibly leaving a sulphate ion and tend to cross link the PPy in an inert atmosphere [54].

Kinetics of PPy textile degradation has been studied as function of dopant ions and temperature of reaction, in our previous study [10].

4.3. γ Ray Irradiation Polymerization

4.3.1. Electrical Conductivity

The electrical conductivity was measured on the surface of the fabrics. The value obtained for surface resistivity is 1.06 X 10^7 ohms/square which is quite high as compared to the Ppy prepared by other methods like chemical oxidation, electrochemical etc. This may be due to the fact that α – β` mode of bonding of PPy molecules. Additionally it may note that there was no oxidant or dopant used during synthesis of PPy.

4.3.2. FITR Spectroscopy

FTIR Studies were carried out by many researchers to understand the mechanism and bonding involved in PPy synthesized through γ irradiation polymerization [50, 68]. According to that, it shows strong absorption bands at around 3389cm^{-1} corresponding to N-H stretching, low intensity peaks at 2927 cm^{-1} and 2814 cm^{-1} corresponding to aromatic C-H stretching vibrations. The band at 1250 cm^{-1} and 1100 cm^{-1} corresponds to breathing vibrations of the Pyrrole ring and at 1310 cm^{-1} attributable to C-N in plane deformation vibration modes[1]. The bond of C-H in plane deformation vibration was situated at 1043 cm^{-1} and of the C-C out of plane ring deformation vibrations or C-H rocking is at 681cm^{-1}. In order to determine the mode of bonding, the bands at 788 cm^{-1}, 730 cm^{-1} and 820 cm^{-1} were analyzed. When α – α` bonding i.e., 2, 5 coupling exists a strong absorption band occurs at 788 cm$^{-1}$. On the other hand when α – β` bonding i.e., 2,4 coupling occurs, absorption band found at 732 cm^{-1} and 825 cm^{-1}. It was indeed observed that when FeCl$_3$ or APS were used as oxidants in chemical polymerization method, peak related only observed at 788 cm^{-1} band corresponding to α – α` bonding. However in the case of γ – radiation samples, strong absorption band was observed at 732 cm$^{-1}$ and a weak absorption band at 825 cm$^{-1}$ was observed, predominantly showing $\alpha - \beta`$ bonding exists with some admixture of $\alpha - \alpha`$ bonding.

Another noteworthy change in the γ – irradiated samples is the absence of absorption bands at 1558 cm^{-1} and 1488 cm^{-1} corresponding to ring vibrations of Pyrrole. Instead a very strong absorption band is observed at 1702 cm^{-1} assigned to C=N bond. Thus breaking of C=C in the ring can lead to formation of C=N. Possibility such rapid changes on γ – radiation may lead to formation of $\alpha - \beta`$ type of polymerization. Absorption band at 1700 cm^{-1} additionally indicates formation of C=O as the irradiation reaction was carried out in air.

ATR spectra of the fabrics, together with that for powdered sample showed that the mechanism of polymerization seems to be bonding through 2, 4 instead of 2, 5 as found with other methods. This is probably due to fact that the interaction of radiations can induce many ions, radicals and radical cations. The electrons produced in the process can recombine with pyrrole monomer and induce additional reactions. Such reactions can initiate polymerization of monomer and polypyrrole is formed [51]. Longer polymer chains are desirable for better electrical conduction. However the chains can also grow on sides, leading to branching and cross linking process which is difficult to avoid. The formation of conjugated bonds and interchain bonding can lead to electron/hole hopping across the chains. If more PPy cations are available, more electron/hole pairs can be formed leading to higher conductivity. The conductivity can rise to yet higher level, depending on the conditions of preparation and doping ions.

During chain propagation whether the chains are formed through 2, 5 bonding of the Pyrrole molecules, a mechanism normally seen, or through 2, 4 bonding is a question of concern. In either case, co planarity of chains and good cross linking can lead to better conjugation and conduction within the plane. However from the point of minimization of energy, helical growth of chain and intermolecular bonding could be a great advantage.

4.3.3. Morphology

It was observed that the morphology of PPy formed through γ-irradation, is mainly globular in nature with a size of globules in the range of 0.1 μm to 0.8μm. The average size of globule is 0.5μm. It seems that although the globules are overlapping one another, yet each globule can be seen separatey as a differeent entity. Additionally it may be noted that most of the particles are of uniform size.

4.4. UV-Induced Radical Polymerization

4.4.1. Electrical Characteristics

The surface resistivity of the PPy fabrics prepared by UV photo irradiation was measured. The value obtained was 1×10^4 ohms/square. This may be because of the polymerization does not give a long chain and therefore the film was brown in colour.

The UV-photo processing can deposit PPy on fabrics as well as other substrates. Although the detailed mechanism of the UV-deposition of PPy is still uncertain, it can be as follows.

It is well known that polymerization occurs when a radical cation or an active radical reacts with more than one monomer before picking up an electron. The first step in a polymerization process is the formation of a radical cation or an active radical. In this case, the initial step may come from two routes: firstly, UV irradiation can cause excitation or charge separation in the oxidant. After excitation, the conduction-band electron (e^-), and the valence-band electron hole, oxidant may undergo either electron transfer reaction with pyrrole monomer to form a radical cation or recombine. Secondly, monomer itself may form an active radical by abstraction of hydrogen from the oxidant under UV irradiation. This radical or radical cation is a reactive species and has a number of possible pathways: (a) It can recombine with an electron to form monomer again; (b) it can react with other species, which are present in the solution, to give side products; (c) it could react with another pyrrole monomer to give a dimer. The next step towards the polymerisation is to produce a dimer radical cation or a dimer radical. The dimer is more readily excited than the monomer because of its more extensive p-orbital system. Finally, such process continues building up longer Ppy chain.

4.4.2. FTIR Spectroscopy

FTIR spctra of PPy formed by UV-induced polymerization were reviewed in depth to analyze chemical structural properties of PPy formed. Based on that, the related peak positions and their assignments were reported as in Table 1.

FTIR characterization review shows a broad absorption band between 3400 and 3000 cm^{-1} corresponding to NH stretching (NH), aromatic CH stretching (CH) and free-carrier absorbance in the doped Ppy. The region below 1700 cm^{-1} shows characteristic Ppy bands. The so-called doping vibration bands can be seen at about 960 and 750 cm^{-1}. The bands due to NH stretching (NH) (3400 cm^{-1}), aromatic (3100 cm^{-1}) and aliphatic (3000–2800 cm^{-1}) CH stretching (CH) were found to be present in the doped Ppy. It is noted that there are weaker peaks in the case of UV-irradiation which is due to the micro structural changes by UV-photo processing. A hump at 780 shows that the mode of bonding is predominating α –α` (2, 5 linkage).

Table 1. Peak positions of FTIR of polypyrrole prepared by UV-irradiation method

Peak Position	Assignments
3213	O-H stretching Vibration
2381	C-H stretching band of aromatic ring
2337	
2323	
1721	C=C ring stretching of pyrrole
1401	C-H vibration/C-N stretching vibrations of ring
1363	C-H bending vibration
1296	C-N bond amines
1243	C=C/C-C stretching vibrations of the pyrrole ring
1184	C-N bond amines, amides
1110	In plane deformation of OH group C-O symmetric stretching
780	2-5 linkage (α-α)
638	C-N bond

4.4.3. Morphology Studies

The morphology of UV induced polymerisation seems to be different than those with other methods. In chemical, electrochemical and gamma polymerisation we observe globular nature. But in UV induced we found grain like morphology. The shapes of grains are random and the sizes of the grains vary from 0.22µm to 0.54µm. The average size of the grain is 0.381 µm.

Polymerization of pyrrole using various methods like (i) Chemical (ii) Electrochemical (iii) γ – irradiation and (iv) UV Photo irradiation has been investigated. The main purpose of using different routes was to study the mechanisms involved in each method and evaluate advantages of each method. Like the Electrochemical method gives PPy in film form and the conductivity is very good. The morphology of polypyrrole prepared by electrochemical method showed globular nature. However the sizes of the globules are large and variation of sizes is over a wide range. The sizes vary from 2 to 30 microns.

4.4.3.1. Comparison of the Methods

Plasma grafting was done for some samples as described earlier. These samples were analysed by all characterization techniques. In Plasma followed by chemical polymerization, it is possible to modify the surface properties of a substrate with plasma treatment, while retaining the bulk properties of the substrate material to deposit PPy coating.

In the γ – irradiation, oxidants can be avoided and therefore the PPy obtained is in highly pure form. Polypyrrole as obtained as globule but had a size in the nanometre range, much less than that obtained by chemical and electrochemical polymerization techniques. The FTIR spectra of γ – irradiation preparations revealed the possibility of α – β` bonding mode on account of rupture of the pyrrole ring accompanied by C=N formation. Uniform and small sized globular morphology probably gives highly adherent PPy films.

Table 2. Comparative study of PPy formation by different techniques

Techniques of polymerization	Dopant used	Peak Position of FTIR and Assignments		SEM Observation	Surface Resistivity (ohms/square)
Chemical Polymerization	Chloride	780	α - α' bonding exists	Globular deposition	50-5000
Plasma followed by chemical polymerization	--	780	α - α' bonding exists	Globular deposition	20 - 3000
Plasma Grafting	-	730 820 1720	Ring opening Planar	Cauliflower shaped having small globules within it.	$5 \times 10^6 - 1 \times 10^7$
Electro chemical Polymerization	PTS NSA	800 790	α - α' bonding exists	Cauliflower shaped having small globules within it.	10-30
Gamma-Irradiation	--	732 794 825	α - β bonding exists with some admixture of α- α' bonding	Nano particle uniform size of PPy globules	1.0×10^7
UV-irradiation	---	780	α- α' bonding exists	Granular deposition	1.0×10^4

Polypyrrole formation on fabric by UV irradiation is a very simple method which can done easily and is environmentally friendly. It is a one-step oxidative polymerization route to fabricate polypyrrole (PPy) on any substrate using hydrogen peroxide as a weak oxidant,

The electrical conductivity of each preparation was found to be different. The conductivity of Polypyrrole prepared from electrochemical polymerization was found to be highest and that of Gamma irradiation was lowest. The Table 2 gives the comparative study of PPy formed by different techniques highlighting the uniqueness of each technique.

5. POLYPYRROLE USE IN SMART TEXTILE APPLICATION

Smart textiles is the area which has its applications in the field of medical, health-care, sports, fitness monitoring, fashion, entertainment, protection, and military. The role of electrical conductivity is paramount in such textiles considering the involvement of tasks such as sensing, actuation and data transmission. Some of the earlier smart textile systems relied on attachment of electronic devices to textile matrix for achieving desired functionality. Gradually, the focus of research has shifted towards replacing hard electronic parts with textiles embedded with electronic functionalities. The convergence of electronics, electrical engineering and textile technologies has the potential to combine the positive attributes of each technology, the speed and computational capacity of modern electronics with the flexible, comfortable and continuous nature of textiles.

All these developments make it possible to make smart biomedical garments, biofeedback systems and smart occupancy detectors which includes the products as wearable motherboard [developed by Georgia Tech (USA)], smart garment WEALTHY [developed bySmartex (Italy))] and Lifeshirt (by Vivometrics Inc). Here the garment itself consists of a grid of optical and electro conductive wires through which data coming from the sensors are transmitted to a processing unit. One of BTRA's developments consists of an occupancy detecting textiles named as "smart mat" [57]. This mat (Figure 17) could be utilized in household or restricted areas for detecting intruder entry. When a person steps on it, a signal is generated. This signal can be wirelessly transported up to a distance of 25 meters to raise an alarm. The same principle can be used for occupancy detection in cars, theatres and other places.

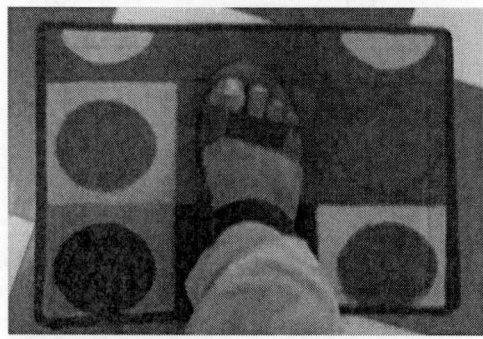

Figure 17. BTRA's developed product-Smart Mat for occupancy detection.

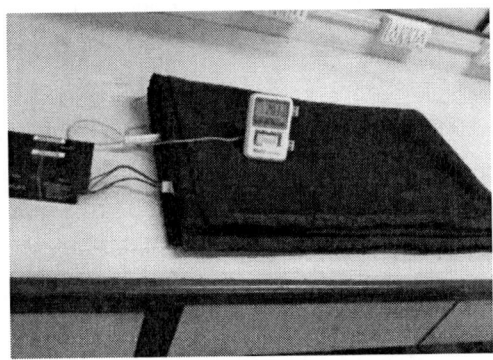

Figure 18. Electrically Heated Blanket incorporated with heat generating textile pads.

BTRA also used CP coated fabrics to achieved temperature rise in heating pads which can be used for incorporation into blankets (Figure 18), jackets, shoes and inner wears to provide warmth and comfort in cold weather conditions. The temperature of the developed pads can be maintained at 40-45°C. A 24 V rechargeable battery is used as a power source. Thesepads can also be used as a thermal therapy which has been used for centuries to combat backaches, muscle and joint pain.

ICP shows promising applications for sensing of gas, humidity [58], temperature and pH [59]. Umana and Waller reported a novel approach to electrode immobilization of an enzyme (glucose oxidase) by electro polymerization of pyrrole [60]. Polypyrrole based aroma sensor was developed by deSouza et al. in 1999, where the change in conductivity with the exposure of volatile gases exploited [61]. PPy films can be employed to construct artificial muscles with tactile sensitivity [62].

Gas sensing behaviour of ICP has been reported in many literatures stating that gases possessing redox properties interact with ICPs by donating and accepting electrons [19, 61-64]. Accordingly, not only gases such as ammonia, carbon dioxide, HCl vapors, but volatile organic compounds (VOCs) have also been reported to be used as analytes. This is important as concepts such as "Electronic Nose (e-nose)" has been devised wherein, e-nose attempts to emulate mammalian nose by using an array of sensors that can simulate mammalian olfactory responses to aromas. The sensors for e-nose are typically based on change in conductivity. These sensing materials are either made from conducting polymer composites, ICPs and metal oxides. Out of this ICPs show a good response to wide range of analytes and have fast response and recovery times especially for polar compounds [61]. Problems related to ICPs include poorly understood signal transduction mechanism, difficulties in resolving some types of analytes, high sensitivity to humidity and the sensor response can drift with time. In spite of such limitations, ICPs offer a prominent advantage in that, unlike metal oxide sensors which require high temperature (350-400°C) for sensing, ICPs can be used at room temperature. ICP based sensors can be utilized for both quantitative and qualitative determination of analytes.

There have been commercial systems of e-nose based on ICPs. The change in conductivity is brought about either through a change in doping level or through a change in polymer conformation. Changes in chain confirmation can also affect conductivity by increasing or decreasing the localization length over which electrons can move freely.

It has been observed that sensing mechanisms of different ways may play part leading to change in conductivity of conductive polymers. These mechanisms are as follows [58]:

1. Oxidize or reduce the polymer, changing number of charge carriers on the polymer chains,
2. Interact with mobile charge carriers on the polymer chain and alter their mobility
3. Interact with dopant molecules,
4. Modify the potential barrier for the hopping process of charge carriers between the polymer chains.
5. May induce the morphological changes e.g., swelling or producing self assemblies of conducting polymer clusters
6. Delocalization of charge or break of double bonds will lead to move charge carriers or reduce the number of charge carriers.

Electron acceptors like NO_2, I_2, O_3, O_2 are able to oxidize partially reduced conductive polymer and therefore increase their doping level [59]. To oxidize conductive polymer, the gas should have a higher electron affinity than conductive polymer. Exposure of SO_2 increase the number of charge carriers in the PPy thus decreasing the resistance [65]. Electron donating gases like NH_3, H_2S reduce and therefore dedope PPy, which leads to an increase in resistance. Ammonia gas was found to decrease the conductivity of PPy coated fabrics, in our previous study [58]. When protonated form of PPy is exposed to ammonia, nitrogen atom of ammonia establishes coordination bond with free atomic orbital of dopant H+. It leads to deprotonation of PPy molecule, resulting in the loss of charge carriers and subsequent decrease in the conductivity.

$$PPyH+ + NH_3 \leftrightarrow PPy + NH_4+$$

Our own studies indicate that, PPy coated fabric shows exactly opposite responses to each other when used for ammonia (increase in resistance) and ethanol gas (decrease in resistance) detection. Ethanol is a good dielectric medium with dielectric constant of 23. Insertion of ethanol between PPy molecules and dopant ions reduces the coulomb interaction, which enhances the hopping rate and hence the conductivity. The response and recovery rate of PPy coated fabric toward ethanol gas was higher as compared to ammonia gas [58].

CONCLUSION

In summary, tuneable conducting property of polypyrrole make it suitable for various applications. This chapter provides valuable information for better understanding of the mechanism involved in synthesis of polypyrrole. Polypyrrole can be synthesized by various methods such as chemical polymerization, electrochemical polymerization, plasma polymerization, γ – irradiation and UV Photo irradiation. Several simple ways can be adopted to achieve uniform conductivity in polypyrrole and many strategies can apply to use it in commercial products. The in situ chemical polymerization of PPy has been successfully carried out by adopting principle of jigger machine, to make flexible conductive fabric.

The resultant fabrics are characterized for electrical conductivity, morphology and fine structure as well as chemical alterations. Electrical surface resistivity value achieved for this chemically synthesized, polypyrrole coated cotton fabrics are in the range of 15 – 5000 Ω/sqr. The FTIR characterization confirmed deposition of PPy onto the fabrics. The SEM characterization demonstrated the morphological features of deposited PPy. The addition of sulfonic acid performs the dual role of improvement in conductivity and atmospheric stability in the studied dopant concentration range.

The electrochemical polymerization of PPy was carried out successfully on the top of PPy-cotton fabric prepared by chemical polymerization. The heat stability of electrochemically prepared fabrics was found to be remarkably superior to only chemical polymerized fabrics. The presence of dopant ions would have a favourable impact in certain application areas. The photo induced polymerization, plasma polymerization and gamma polymerization also gives uniform coating of PPy, but the conductivity achieved is comparatively lower than that of by electrochemical and chemical polymerization.

The developed e-textiles based on PPy coated fabrics were found to be suitable for various applications in the areas of smart textiles.

REFERENCES

[1] Perumalraj R., Tholkappiyan E., Kanimozhi J., Jayashree P. and Ramya R, 2011, 'Electrostatic discharge measurement for conductive textiles materials,' *Asian Textile Journal*, 58-64.

[2] Onar N., Aksit, A. C., Ebeoglogil, M. F., Birlik, I. & Celik, E., 2009, Structural, electrical, and electromagnetic properties of cotton fabrics coated with polyaniline and polypyrrole, *Journal of Applied Polymer Science*, 114, 2003-2010.

[3] Varesano A., Aluigi A., Florio L. and Fabris R., 2009, Multifunctional Cotton Fabrics, *Synthetic Metals*, 159, 1082-1089.

[4] Kaynak A., Kakansson E. and Amiet A., 2009, The influence of polymerization time and dopant concentration on the absorption of microwave radiation in conducting polypyrrole coated textiles, *Synthetic Metals*, 159, 1373-1380.

[5] Romero E., Molina J., Rio A., Bonastre J. and Cases F., 2011, Synthesis of PPy/PW12 O3-40 organic-inorganic hybrid material on polyester yarns and subsequent weaving to obtain conductive fabrics, *Textile Research Journal*, 81(14), 1427-1437.

[6] Shang S., Yang X., Tao X. and Lam S., 2009, Vapor-phase polymerization of pyrrole on flexible substrate at low temperature and its application in heat generation, *Polymer International*, DOI 10.1002/pi.2709.

[7] Bhat, N. V., Seshadri, D. T., & Radhakrishnan S. 2004, Preparation, Characterization, and Performance of Conductive Fabrics: Cotton + PANi, *Textile Research Journal*, 2004, 74, 155-166.

[8] Esfandiari A., 2008, PPy Covered Cellulosic and Protein Fibres Using Novel Covering Methods to Improve the Electrical Property, *World Applied Sciences Journal* 3(3), 470-475.

[9] Wallace G. G., Spinks, G. M., Kane-Maguire, L. A. P., and Teasdale P. R., 2003, Synthesis and structure of polyanilines, in *Conductive Electroactive Polymers*, (CRC Press), 129.

[10] Patil A. J. and Deogaonkar S. C., 2012, Conductivity and Atmospheric aging studies of Polypyrrole coated cotton fabrics, *Journal of Applied Polymer Science*, Volume 125, Issue 2, pages 844–851.

[11] Deogaonkar S. C. and Patil A. J., 2014, Development of conductive cotton fabric by in situ chemical polymerization of Pyrrole using ammonium peroxidisulphate as oxidant, Indian *Journal of Fibre & Textile Research,* Volume 39, 35-138.

[12] Ferrero F., Napoli L., Tonin C., Varesano A., 2006, Pyrrole chemical polymerization on textiles: Kinetics and operating conditions, *Journal of Appiedl Polymer Science*, 102, 4121-4126.

[13] Rasika Dias H. V., Fianchini M, Gamini Rajapakse R. M., 2006, Greener method for high quality polypyrrole, *Polymer*, 47, 7349-7354.
[14] Neoh K. G., Kang E. T., Tan T. C., 1989, Effects of acceptor level on chemically synthesized polypyrrole-halogen complexes, *Journal of Applied Polymer Science*, Vol 37, Issue 8, 2169-2180.
[15] Chao T. H. and March J., 1988, A study of Polypyrrole synthesized with oxidative transition metal ions, *Journal of Applied Polymer Science*, 26, 743-753.
[16] Lei J, Cai Z, Martin C. R. 1992, Effect of reagent concentrations used to synthesize polypyrrole on the chemical characteristics and optical and electronic properties of the resulting polymerSynth. *Metals*, 46, 53-69.
[17] Saville P., *Polypyrrole- Formation and Use, DRDC*, Atlantic TM 2005-004.
[18] Patil A. J. and Pandey A, 2012, A novel approach for in-situ polymerization of polypyrrole on cotton substrates, *Indian Journal of Fibre &Textile Research*, Vol 37(2), 107-113.
[19] Chitte H. K., Bhat, N. V., Walunj, V. E., & Shinde, G. N., 2011, Synthesis of Polypyrrole Using Ferric Chloride ($FeCl_3$) as Oxidant Together with Some Dopants for Use in Gas Sensors, *Journal of Sensor Technology*, 1, 47-56.
[20] Liu C., Cai Z., Zhao Y., Zhao H., Ge F., 2006, Potentiostatically synthesized flexible polypyrrole/multi-wall carbon nanotube/cotton fabric electrodes for supercapacitors. *Cellulose*, Vol. 23, Issue-1, 637-648.
[21] Sak-Bosnar M., Budimir M. V., Kovac S., Kukulj D. and Duic L., 1992, Chemical and electrochemical characterization of chemically synthesized conducting polypyrrole, *Journal of Polym Sci. Part A:Polym Chemistry*, 30, 1609-1614.
[22] Xie Y., Z. Qiao, M. Chen, X. Liu, Y. Qian, 1999, 'γ-Irradiation route to semiconductor/polymer noncable fabrication,' *Adv. Mater.* 11, 1512-1515.

[23] Milojković S., D. Kostoski, J. Čomor, J. M. Nedeljković, 1997, Radiation induced synthesis of molecularly imprinted polymers, *Polymer* 38(11), 2853.

[24] Seguchi T., T. Yagi, S. Ishikawa, Y. Sano, 2002, New material synthesis by radiation processing at high temperature—polymer modification with improved irradiation technology, *Radiat. Phys. Chem.* 63(2002)35-40.

[25] Zhang X., M. Wang, T. Wu, S. Jiang, Z. Wang, J. 2004, In Situ Gamma Ray-Initiated Polymerization To Stabilize Surface Micelles *Am. Chem. Soc.* 126,(2004)6572.

[26] Tasdelen M. A. and Yagci Y. 2011, 'Photochemical Methods for the Preparation of Complex Linear and Cross-linked Macromolecular Structure,' *Aust. J. Chem.* 64, 982–991.

[27] Hageman H. J., 1985, Photo initiators for free radical polymerization, *Progress in organic coatings*, 13, 123-150.

[28] Gruber H. F., 1992, Photo initiators for free radical polymerization, *Progress in Polymer Science*, 17, 953-1044.

[29] Yagci Y., Jockusch S., Turro N. J., 2010, 'Photo initiated Polymerization: Advances, Challenges, and Opportunities,' *Macromolecules*, 43, 6245-6260.

[30] Eliasson B., U. Kogelschatz, 1988, UV excimer radiation from dielectric-barrier discharges *Applied Physics B*, 46, 299-303.

[31] Kogelschatz U., 1992, Silent-discharge driven excimer UV sources and their applications, *Appl. Surf. Sci.* 54, 410-423.

[32] Esrom H., J. Y. Zhang, U. Kogelschatz, 1995 New approach of a laser-induced forward transfer for deposition of patterned thin metal films, *Applied Surface Scienec*, 86, 202-207.

[33] Kanazawa, K. K., A. F. Diaz, W. D. Gill, P. M. Grant, G. B. Street, G. P. Gardini, J. F. Kwak, 1980, Polypyrrole: An electrochemically synthesized conducting organic polymer *Synth. Met.*, 1, 329-336.

[34] Wang L., Li X., Yang Y., 2001, Preparation, properties and applications of polypyrroles. *Reactive and Functional Polymers*, 47, 125-139.

[35] Khare S., Deogaonkar S. and Savadekar N., 2015, Polypyrrole coated Nonwoven substrate for electromagnetic shielding, *AATCC Journal of Research,* Volume 2, No.1Pages 11-15.
[36] Nicolau Y. F., Davied S, Genoud F, Nechtschein M, Travers J. P. 1991, Polyaniline, polypyrrole, poly(3-methylthiophene) and polybithiophene layer-by-layer deposited thin films, *Synth. Met.* 42, 1491-1494.
[37] Khedkar S. P. and Radhakrishnan S., 1997, Long-term aging and stability of conductivity in vapour phase deposited polypyrrole films, *Polymer Degradation and Stability*, 57, 51-58.
[38] Kuhn H. H., 1997, Adsorption at the liquid/solid interface: conductive textiles based on polypyrrole' *Textile Chemist and Colourist*, 29, 17-21.
[39] Zoppi R. A., De Paoli, M. A. 1996, Chemical preparation of conductive elastomeric blends: polypyrrole/EPDM—II. Utilization of matrices containing cross linking agents, reinforcement fillers and stabilizers, *Polymer,* 37, 1999-2009.
[40] Garg S., Hurren C. & Kaynak A. 2007, Improvement of adhesion of conductive polypyrrole coating on wool and polyester fabrics using atmospheric plasma treatment, *Synthetic Metals*, 157-1, p. 41-47.
[41] Bhattacharya A., De S., Bhattacharya S. N. and Das S, J., 1994, Transport properties of $FeCl_3$-doped polypyrroles at different dopant concentrations, *Phys. Cond. Matter* 6 10499.
[42] Su S. J. and Kuramoto N., 2000, Processable polyaniline–titanium dioxide nanocomposites: effect of titanium dioxide on the conductivity *Synth. Met.* 114, 147-153.
[43] Vishnuvardhan T. K., Kulkarni V. R., Basavaraja C. and Raghavendra S C, 2006, Synthesis, characterization and a.c. conductivity of polypyrrole/Y2O3 composites *Mater. Sci.,* 29(1), 2006, 77-83.
[44] Chen W., Xingwei L., Gi X., Zhaoquang W. and Wenquing Z., 2003, Magnetic and conducting particles: preparation of polypyrrole layer on Fe_3O_4 nanospheres, *Appl. Surf. Sci.* 218, 216.

[45] Chen X. B., Issi J. P., Devaux J, Billaud D., 1997, The stability of polypyrrole and its composites *Journal of Material Sci*, 32, 1515-1518.

[46] Troung V. T., Codd A. R., Forsyth M, 1994, Dielectric properties of conducting polymer composites at microwave frequencies, *Journal of Material Sci.*, 29, 4331-4338.

[47] Stanke D., Hallensleben M. L., Toppare L., 1995, Oxidative polymerization of pyrrole with iron chloride in nitromethane *Syntheic Metals*, 72,159-165.

[48] Dams R., Vangeneugden D. and Vanderzande D., 2013, 'Atmospheric Pressure Plasma Polymerization of In Situ Doped Polypyrrole,' *The Open Plasma Physics Journal*, 6, 7-13.

[49] Gangopadhyay R. and De A., 2000, Conducting Polymer Nanocomposites: A Brief Overview, *Chemistry of Materials*, 12, 3, 608-622.

[50] Karim M. R., Lee C. J., Lee M. S., 2007, 'Synthesis of conducting polypyrrole by radiolysis polymerization method,' *Polymers for Advance Technologies*, 18, 916-920.

[51] Leonavicius K., Ramanaviciene A. and Ramanavicius A., 2011, 'Polymerization Model for Hydrogen Peroxide Initiated Synthesis of Polypyrrole Nanoparticles,' *Langmuir* 27, 17, 10970-10976.

[52] Deogaonkar S. C., 2018, 'Improvement of polypyrrole coating adhesion on polyester fabric by atmospheric pressure plasma technology' *Colourage*, 29-35.

[53] Kuhn, H. H., Child, A. D., Kimbrell, W. C., 1995, Toward real applications of conductive polymers, *Synthetic Metals*, 71, 2139-2142.

[54] Kassim A., Basar Z., Mahmud H. N. M., 2002, Effects of preparation temperature on the conductivity of polypyrrole conducting polymer, *Proc. Indian Acad. Sci. (Chem. Sci.)* Vol. 114(2), 155-162.

[55] Forsyth M, Smith M. E., 1993, Solid state NMR characterization of polyrrole: The nature of the dopant in PF_6^- doped films, *Synthetic Metal*, 55, 714-719.

[56] Leonavicius K, Ramanaviciene A and Ramanavicius A, 2011, 'Polymerization Model for Hydrogen Peroxide Initiated Synthesis of Polypyrrole Nanoparticles,' *Langmuir* 27, 17, 10970-10976.
[57] Deogaonkar S. C., 2015, 'Conductive textiles and its applications' paper presented at *56th Joint Technological Conference of ATIRA, BTRA, NITRA and SITRA at SITRA*, Coimbatore on 29th December 2015.
[58] Deogaonkar S. C. and Bhat N. V., 2015, 'Polymer based fabrics as transducers in ammonia and ethanol gas sensing,' *Fibres and Polymers*, Volume 16, No. 8, 1803 – 1811.
[59] Lange U., V. Nataliya, Roznyatovskaya V., Mirsky V. M., 2008, 'Conducting polymers in chemical sensors and array,' *Analytica Chimica Acta* 614, 1-26.
[60] Umana M and Waller J, 1986, Protein-modified electrodes. The glucose oxidase/polypyrrole system, *Analytical Chemistry*, 58(14), 2979-2983.
[61] De-Souza J. E. G., Neto B. B., dos Santos F. L., de Melo C. P., Santos M. S., Ludermir T. B., 1999, Polypyrrole based Aroma sensor, *Synthetic Metals*, 102, 1296-1299.
[62] Otero T. F. and Cortes M. T., 2003, Artificial Muscles with Tactile Sensitivity, *Advanced Material*, 15, 279-282.
[63] Dhawale D. S., Dubal D. P., More A. M., Gujar T. P., Lokhande C. D., 2010, Room temperature liquefied petroleum gas (LPG) sensor, *Sensors and Actuators B: Chemical*, 147, 488-494.
[64] Kharat H. J., K. P. Kakde, P. A. Savle, K. Datta, P. Ghosh and M. D. Shrisat, 2007, Synthesis of polypyrrole films for the development of ammonia sensor, *Polym. Adv. Tech.*, 18, 397-402.
[65] Prissanaroon W., Ruangchuay L, Sirivat A, Schwank J, 2000, 'Electrical conductivity response of dodecylbenzene sulfonic acid-doped polypyrrole films to SO_2–N_2 mixtures' *Syntheic Metals*, 114, 65-72.
[66] Mitchell G. R. and Geri A. 1987, Molecular organisation of electrochemically prepared conducting polypyrrole films, *Journal of Physics D: Applied Physics*, 20, 1346.

[67] Davidson R. G., Hammond L. C., Turner T. G. and Wilson A. R., 1996, An electron and X-ray diffraction study of conducting polypyrrole/dodecyl sulphate, *Synth. Met.*, 1996, 81:1, 1-4.

[68] Cui Z., Coletta C., Dazzi A., Lefrancois P., Gervais M., Neron S. and Remita S., 2014, Radiolytic Method as a Novel Approach for the Synthesis of Nanostructured Conducting Polypyrrole *Langmuir,* 30, 46, 14086-14094.

[69] Chougule M. A., Pawar S. G., Godse P. R., Mulik R. N., Sen Shaswati, Patil V. B., 2011, Synthesis and Characterization of Polypyrrole (PPy) Thin Films, *Soft Nanoscience Letter*, 1, 6-10.

In: Pyrrole: Synthesis and Applications
Editor: Colin Welch

ISBN: 978-1-53617-137-2
© 2020 Nova Science Publishers, Inc.

Chapter 2

RECENT RESEARCH ADVANCES IN THE AQUEOUS PHASE SYNTHESIS OF PYRROLES

Venkata Durga Nageswar Yadavalli[1,*], *PhD,*
Swapna Kokkirala[1], *PhD*
and Venkanna Avudoddi[2], *PhD*

[1]Indian Institute of Chemical Technology,
Tarnaka, Hyderabad, India
[2]Osmania University, Telangana, India

ABSTRACT

Among the heterocyclic systems, pyrrole skeleton is widely distributed in many natural and biologically relevant molecules such as alkaloids, porphyrins, and co-enzymes. Pyrrole derivatives play a crucial role in organic chemistry, medicinal chemistry, and heterocyclic chemistry. Pyrrole scaffold is extensively used in the synthesis of drug molecules with various pharmacological properties as well as in material sciences. Several synthetic protocols have been reported by enthusiastic researchers worldwide during the last decade.

---

[*] Corresponding Author's Email: dryvdnageswar@gmail.com.

In view of several environmental and regulatory issues, promoting sustainable green synthetic processes has become a global necessity. This paradigm shift from classical organic reactions to eco-friendly reactions as well as technologies has led to the evolution of green chemistry. However, for the development of environmentally benign reactions, the selection of alternate nontoxic solvent medium plays a significant role.

Water, a universal solvent, which is an essential element of life on our planet, is involved in different biological processes in nature. Despite its potential role in nature, due to the complexity of present-day organic reactions and the poor solubility of organic compounds in water, its role as a sole solvent in conducting organic reactions is limited. However, several researchers, with an interest to develop environmentally acceptable reactions, have reported aqueous phase reactions. In view of immense attention shown globally towards water medium reactions, an attempt is made to review the scientific research reports pertaining to the "Recent Research Advances in the Aqueous Phase Synthesis of Pyrroles" during the last decade.

INTRODUCTION

Pyrrole, a five-membered aromatic heterocyclic compound is a well known biologically active scaffold, and its derivatives occur widely in nature. Chlorophyll, bile pigments, and heme derivatives are a few prominent examples of natural pyrroles. Pyrrole structural motif represents itself in many biologically active molecules such as alkaloids, coenzymes, and porphyrins. Pyrrole alkaloids have been isolated both from marine and non-marine organisms. Lamellarins, isolated from marine invertebrates, exhibit anti-HIV, and anti-tumor activities [1]. The prodiginines, tripyrrolic alkaloids, exhibit many interesting biological activities [2, 3]. Pyrrole containing compounds display remarkable pharmacological activities such as anti-bacterial, anti-viral, anti-inflammatory, anti-tumoral, and antioxidant activities. Non-steroidal anti-inflammatory zomepirac and tolmetin, anthelmintic pyrivinium, renal cancer drug sunitinib, and cholesterol-lowering agent atorvastatin are some of the prominent marketed drugs containing pyrrole skeleton (Figure 1). Pyrrole synthesis is frequently used in the synthesis of agrochemicals, dyes, pharmaceutical, and photographic chemicals. Moreover, they are also essential intermediates in the synthesis of heterocycles, [4] natural products as well as chemicals used in material science [5].

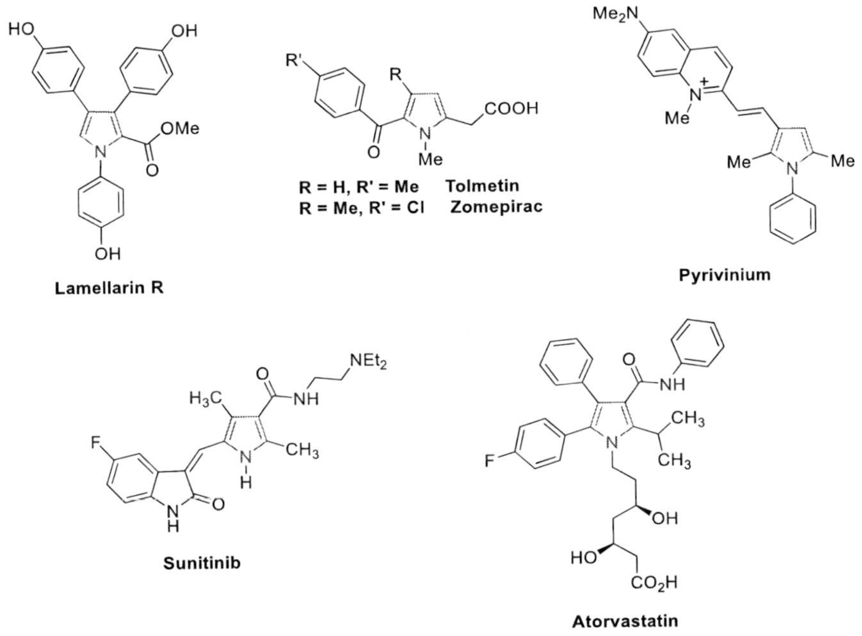

Figure 1. Some Pyrrole derived drugs.

Among pyrroles, tetrasubstituted pyrroles are known for exhibiting anti-bacterial, anti-viral, antioxidant, and anti-convulsant activities [6]. Consequent to the broad range of activities exhibited by the compounds with pyrrole moiety, researchers are continuously attracted towards design and development of new strategies for the synthesis of pyrroles.

In general, organic synthetic processes are carried out in a solvent medium, which is commonly toxic, inflammable, and hazardous. Moreover, experimental laboratory procedures involve extraction, evaporation, and spillages of organic solvents, leading to environmental pollution. Industrial use of solvents and other chemicals cause groundwater contamination in the form of effluents. In view of the serious hazards posed by the use of a wide variety of volatile, nonvolatile, toxic, inflammable, nonpolar as well as polar aprotic solvents, design and development of eco-friendly, sustainable reaction processes involving nonconventional alternate solvent media are increasingly being practiced by researchers across the world. This paradigm shifts from classical methods to green chemical approaches led to a surge in the number of potential publications in this particular area of research.

Among alternate media, a natural, widely available, non-toxic, nonflammable solvent-water is a better choice for carrying out chemical reactions. Advantages of water medium reactions are operational simplicity, cost reduction, and eco-friendly nature. Because of this, aqueous phase reactions have attracted the attention of chemists the world over.

The present review is devoted to discussing water medium reactions involving preparation of pyrrole compounds, covering the literature published in the last decade.

RECENT LITERATURE

A basic functionalized and reusable ionic liquid 1-butyl-3-methylimidazolium hydroxide, [bmim]OH catalyzed three-component condensation between acid chlorides, amino acids and dialkyl acetylenedicarboxylates in water, was explained by Issa Yavari and Elaheh Kowsari [7] resulting in functionalized pyrroles (Figure 2). It is assumed that intermediate produced from the reaction of acid chlorides and amino acid on decarboxylation in the presence of [bmim]OH leads to another intermediate which on reacting with dialkyl acetylenedicarboxylate produces another intermediate. This on cyclization and elimination of water yields the end product.

The advantages of this one-pot three-component methodology, as claimed by authors, are a simple experimental procedure, use of cheap and recyclable benign solvent and its adaptability to prepare diversified pyrrole derivatives.

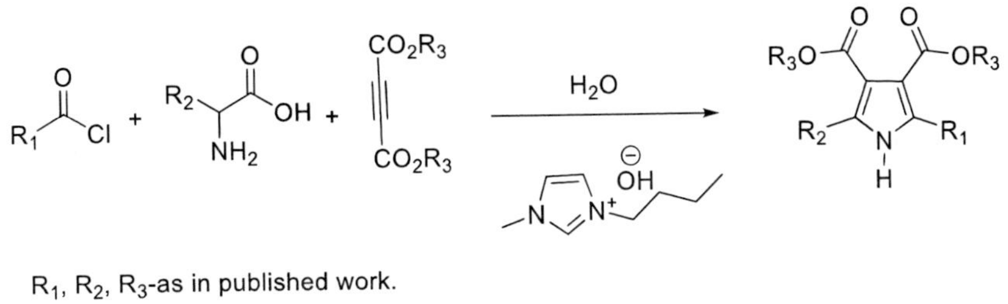

CO_2R_3
H_2O
R_3O
OR_3
R_1
Cl
R_2
OH
NH_2
CO_2R_3
OH
N
N+
R_2
R_1
N
H
R_1, R_2, R_3-as in published work.

Figure 2. Synthesis of highly functionalized pyrrole dervatives.

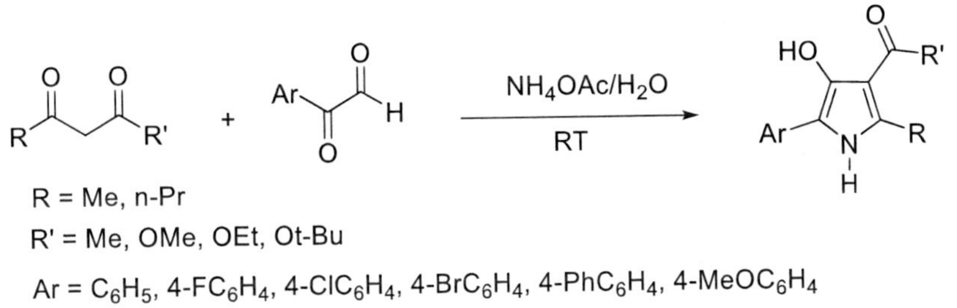

HO
R'
Ar
H
NH_4OAc/H_2O
RT
R
R'
Ar
R
N
H
R = Me, n-Pr
R' = Me, OMe, OEt, Ot-Bu
Ar = C_6H_5, 4-FC_6H_4, 4-ClC_6H_4, 4-BrC_6H_4, 4-PhC_6H_4, 4-$MeOC_6H_4$

Figure 3. Synthesis of 2-alkyl-5-aryl-(1H)-pyrrole-4-ol derivatives.

In a research paper submitted by Behzad Khalili et al. [8] various new 2-alkyl-5-aryl-(1H)-pyrrole-4-ol derivatives were reportedly prepared via a three-component reaction of arylglyoxals with β-dicarbonyl compounds in the presence of ammonium acetate in aqueous medium at room temperature (Figure 3).

As discussed by the authors of this eco-friendly approach to pyrroles, the mechanism involved the attack of enolates on to the phenylglyoxal followed by the condensation of in situ generated 3-hydroxy-1,4-dicarbonyl compound to afford the product in the presence of NH_4^+ by the loss of two water molecules. Authors demonstrated a simple room temperature procedure for the preparation of highly substituted pyrroles from easily available reactants without the presence of any acid catalyst.

In an interesting research work published by Masahiro Yoshida et al. [9], the authors achieved the synthesis of substituted furans as well as pyrroles by platinum-catalyzed cyclizations of propargylic oxiranes and aziridines in water medium (Figure 4). Different propargylic oxiranes were prepared by the epoxidation of the corresponding enynes under basic reaction conditions employing *m*-chloroperbenzoic acid.

Wide range of substituted propargylic oxiranes was finally converted to furan derivatives by platinum-catalyzed cyclization. Exclusive studies were undertaken towards the scope and mechanism of the approach. Similarly, the reaction of various substituted propargylic aziridines afforded corresponding pyrroles. Catalyst screening experiments were carried out to find the most suitable and highly efficient catalyst for these cyclization reactions. As suggested by the authors, the mechanism involved the coordination of platinum to the C≡C bond, followed by the attack of aziridine nitrogen on the alkyne affording the cyclized intermediate. This intermediate on aromatization formed the pyrrolyl platinum species, which on proto-demetalation yielded the pyrrole structure. The methodology provided a broad range of pyrroles as well as furans under aqueous conditions.

Issa Yavari and Elaheh Kowsari [10] described an eco-friendly reaction between dialkyl acetylenedicarboxylates, acid chlorides and amino acids promoted by task-specific basic ionic liquids as catalysts in a water medium, leading to tetrasubstituted pyrroles as end products (Figure 5). Basic ionic liquids have advantages such as stability in water and air, easy separation and recyclability. They also exhibit potential to replace conventional basic catalyst as they are noncorrosive, nonvolatile flexible, and immiscible in organic solvents.

Authors selected benzyl chloride, L-phenyl alanine, and dimethyl acetylenedicarboxylate as model substrates and observed water as an ideal solvent. Extensive studies on the catalytic efficiency of various ionic liquids indicated that [bmin]OH was vital for the reaction. It was also observed that anions had a greater effect than the side chain of the imidazolium cation on the activities of ionic ligands. The general validity of the protocol was investigated to extend to include various acid chlorides and amino acids. Possible mechanistic aspects were also discussed and reported.

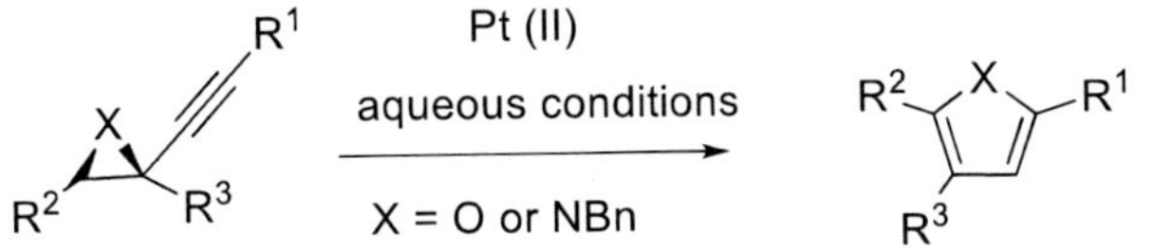
Pt (II)
aqueous conditions
X = O or NBn
R1
R2
R3
X

Figure 4. Synthesis of substituted pyrroles.

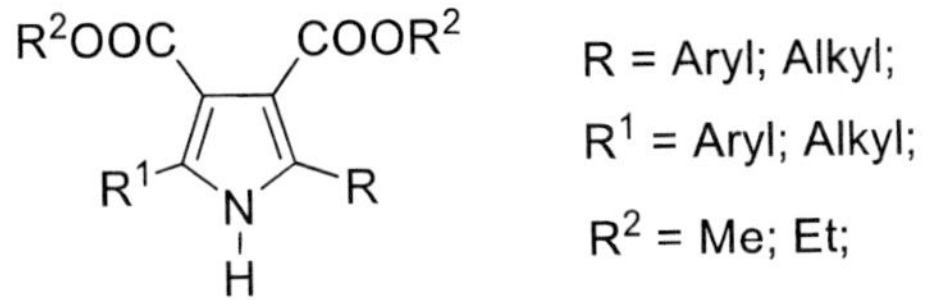
R2OOC
COOR2
R1
R
N
H
R = Aryl; Alkyl;
R1 = Aryl; Alkyl;
R2 = Me; Et;

Figure 5. Tetra substituted pyrroles.

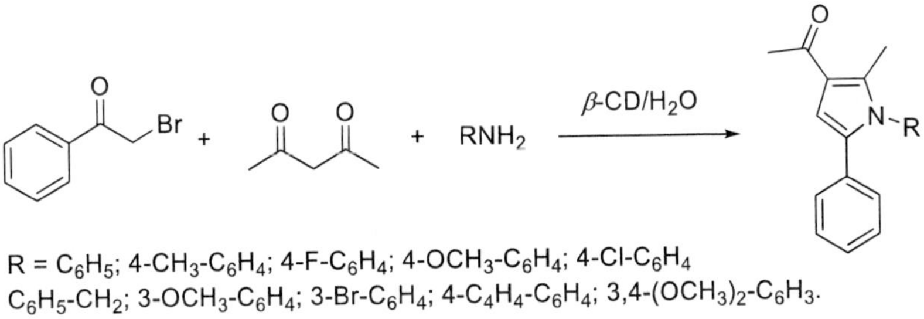
O
Br
+
O O
+
RNH2
β-CD/H2O
N-R
R = C6H5; 4-CH3-C6H4; 4-F-C6H4; 4-OCH3-C6H4; 4-Cl-C6H4
C6H5-CH2; 3-OCH3-C6H4; 3-Br-C6H4; 4-C4H4-C6H4; 3,4-(OCH3)2-C6H3.

Figure 6. Synthesis of N-substituted pyrroles.

Multicomponent reactions are excellent strategies, being employed in the organic transformations. MCRs are convergent in contrast to divergent multi-step synthesis, as the formation of several bonds occurs in one single step. These offer a wide range of advantages avoiding complicated processes and saving large volume of solvents and reagents. There is a tremendous surge of interest and development in three and four-component reactions due to their synthetic efficiency.

S. N. Murthy et al. [11] developed a multi-component approach towards the synthesis of substituted pyrroles utilizing supramolecular catalysis by β-cyclodextrin in water under neutral conditions (Figure 6). β-Cyclodextrins promote the reaction by the reversible formation of host-guest complexes by non-covalent bonding. Activated phenacyl bromide reacts with pentane-2,4-dione and amine to provide pyrrole derivatives.

β-Cyclodextrins are recyclable and reusable at the end of the reaction. This versatile protocol is extended to prepare a wide range of N-substituted pyrroles.

Scope and limitations of microwave-assisted Clauson-Kass pyrrole synthesis under ecofriendly conditions using either water or acetic acid as a medium in the absence of promoters were discussed by authors Kelsey. C. Miles et al. [12] in their research paper (Figure 7). As they claim, the reaction was successful for all common nitrogen reactants when acetic acid was used as a solvent whereas benzylamine and benzamide are resistant to cyclization in the aqueous phase.

Synthesis of N-substituted pyrroles by the reaction of aryl/alkyl, sulfonyl and acyl amines with 2,5-dimethoxy tetrahydrofuran under efficient catalysis of $FeCl_3.7H_2O$ in water medium was reported by Najmedin Azizi and coworkers. [13] After initial studies on a model compound for optimization of the reaction conditions, the scope and limitation of this convenient and straightforward protocol were examined to extend successfully to sterically, electronically and functionally diversified amines in the presence of 2 mol% $FeCl_3.7H_2O$ in water at 60^0C (Figure 8).

Magnus Rueping and Alejandro Parra [14] reported a metal-free, mild, and facile synthesis of tri-substituted pyrroles by reacting enamines with β-bromo nitrostyrenes in water (Figure 9). In this new eco-friendly approach, β-bromo nitrostyrenes were used as tri-functional synthons.

Anna N. Kolontsova et al. [15] described a novel one-pot multicomponent synthesis of 2-amino pyrroles by the reaction of isocyanides with thiophenols and gem-diactivated olefins. The reaction mechanism was discussed by the authors in the research paper (Figure 10).

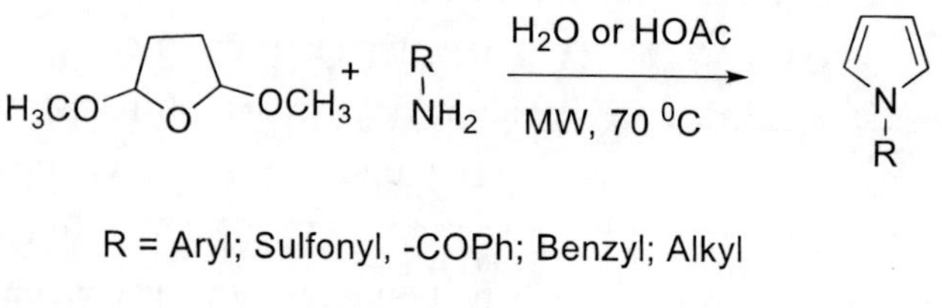

R = Aryl; Sulfonyl, -COPh; Benzyl; Alkyl

Figure 7. Synthesis of N-substituted pyrrole derivatives.

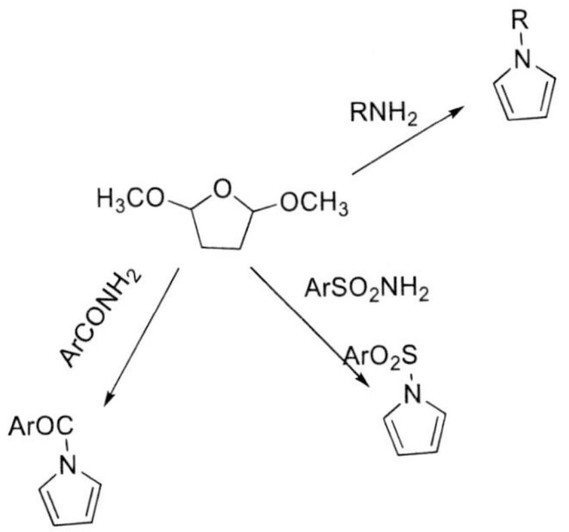

Figure 8. Synthesis of N-substituted pyrroles.

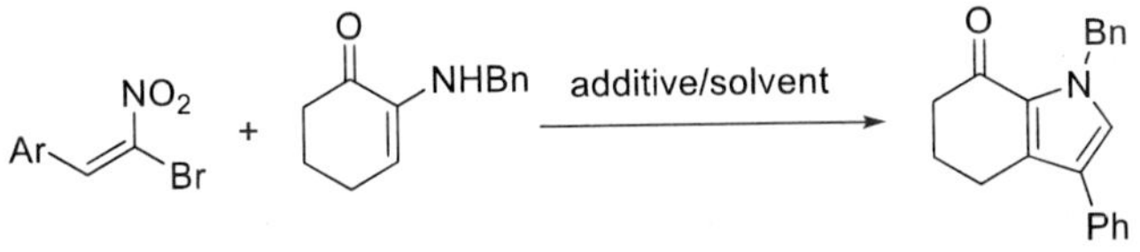

Figure 9. Synthesis of different N-substituted pyrroles.

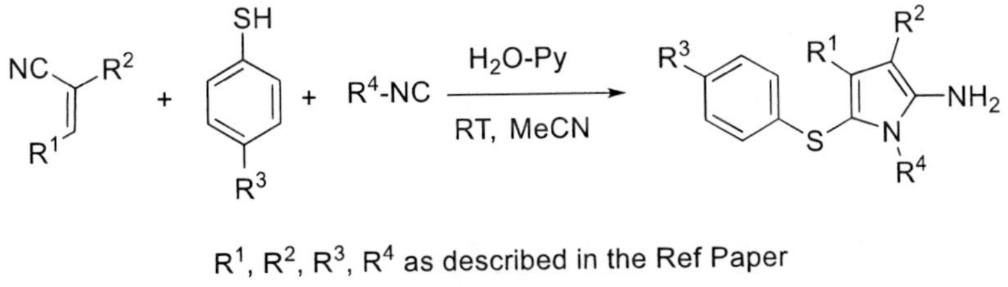

Figure 10. Synthesis of tetra substituted pyrrole derivatives.

Preliminary studies were conducted on benzyl enaminone. Further different parameters like a solvent, reaction time and additives were evaluated, and the scope of the reaction was extended employing several β-bromo nitrostyrenes. It was observed that additives like potassium acetate and other bases complicated the reaction by giving complex mixtures. The generality and efficiency of the reaction were also checked using other cyclic as well as acyclic enaminones. Authors presented the possible mechanism behind this protocol. As explained by them, the first step of the domino reaction involves a conjugate addition of enaminone to a nitrostyrene to give the intermediate, which undergoes subsequent protonation, tautomerization followed by intramolecular nucleophilic substitution and elimination of nitro group to afford expected pyrrole products.

S. Narayana Murthy and Y. V. D. Nageswar, [16] described a mild, environmentally benign and efficient synthesis of diversely substituted N-benzyl pyrroles as well as N-protected methyl pyrrole-2-carboxylates using o-iodoxy benzoic acid, a versatile hypervalent iodine reagent mediated by β-cyclodextrin in aqueous medium at room temperature (Figure 11). Scope of the reaction was extensively investigated.

A convenient synthesis of C-substituted and N-substituted pyrroles in water medium was developed by the reaction of phenacyl bromides, pentane-2,4-dione, and amines using DABCO as reported by H. M. Meshram and coworkers. [17] After establishing the initial three-component reaction, the molar ratio of DABCO was studied until suitable results were obtained (Figure 12). The application of this approach was extended to substituted phenacyl bromides as well as various alkyl, aryl, and benzyl amines to demonstrate the generality. This eco-friendly method also avoids the use of metals or expensive reagents. The possible mechanism, as suggested by authors, involves initial reaction of pentanedione with an amine to form unsaturated aminoketone and tautomerization to form intermediate, which further reacts with the quaternary salt formed between DABCO and phenacyl bromide. The adduct on internal cyclization and dehydration results into the expected product.

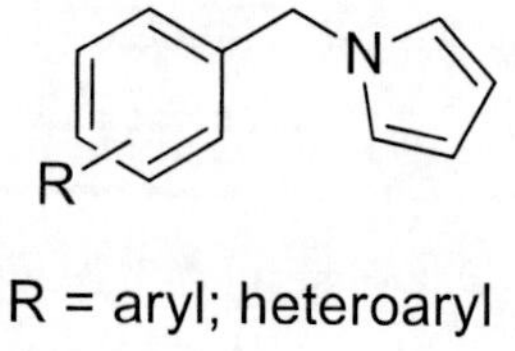

R = aryl; heteroaryl

Figure 11. Diversely substituted N-benzyl pyrroles.

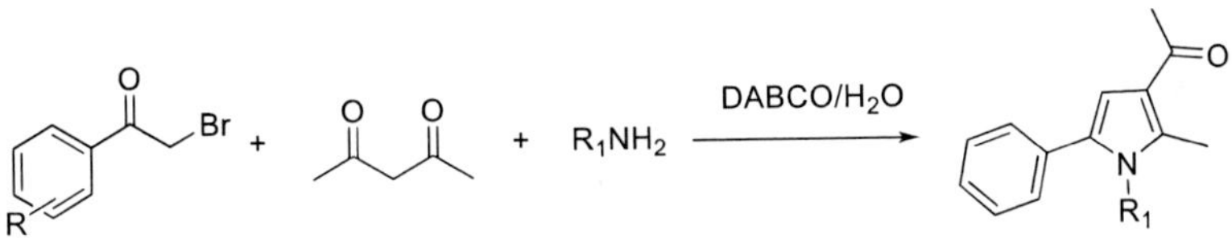

Figure 12. Synthesis of C-substituted and N-substituted pyrroles.

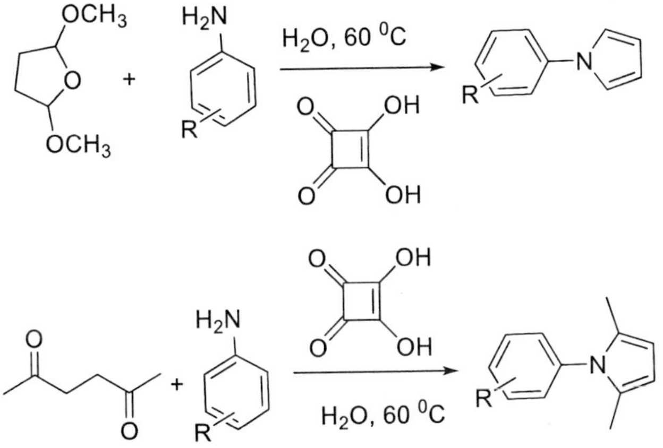

Figure 13. Synthesis of N-substituted pyrrole derivatives.

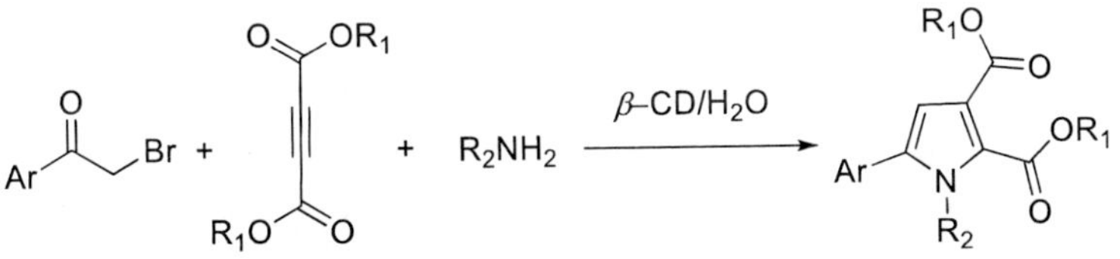

Ar = Ph; 4-Br-C$_6$H$_4$; 4-NO$_2$-C$_6$H$_4$

R$_1$ = Me, Et; R$_2$ = Ph; 4-Me-C$_6$H$_4$; 4-F-C$_6$H$_4$

Figure 14. Synthesis of highly substituted pyrrole derivatives.

Najmadin Azizi and coworkers [18] in an eco-friendly and operationally simple synthetic protocol to N-substituted pyrroles described the reaction of 2,5-dimethoxy tetrahydrofuran and 2,5-hexanedione with diverse aryl amines in green reaction media such as water, deep eutectic solvent and polyethylene glycol under thermal conditions as well as ultrasound irradiation (Figure 13).

Initially, the authors focused on the screening of efficiency of different catalysts for the reaction mentioned above. After the optimization studies, the reaction was reportedly best carried out by utilizing reusable, novel organocatalyst squaric acid. Reactions were successfully conducted between 2,5-dimethoxy tetrahydrofuran and various commercially available aromatic amines having electron-donating as well as electron-withdrawing groups. As reported, Brønsted acidity of squaric acid was the main factor influencing the reactivity as well as selectivity of the process. The key aspects of the mechanism as envisaged by authors are (i) reversible acid-base reaction of aniline with squaric acid (ii) hydrolysis of 2,5-dimethoxy tetrahydrofuran to 1,4-dicarbonyl compound in water (iii) finally condensation of the activated 1,4-dicarbonyl compound with aniline.

Furthermore, the scope and generality of the present protocol were extended to 2,5-hexanedione, and corresponding substituted pyrroles were produced in good to excellent yields.

A novel and efficient synthesis of highly substituted pyrroles was achieved in aqueous medium *via* a multicomponent strategy, by K. Ramesh et al. in their interesting research report [19], the reaction between aromatic amines, phenacyl bromide, and diethyl/dimethyl acetylenedicarboxylate was conducted in the presence of non-toxic, eco-friendly, easily available supramolecular catalyst β-cyclodextrin in water (Figure 14). Cyclodextrins are well known cyclic oligosaccharides with hydrophobic cavities and exhibit host-guest phenomenon. These activate organic molecules and are capable of promoting the reactions in the aqueous phase. As a part of developing green chemical approaches for heterocyclic synthesis, the authors initially conducted a model-controlled experiment using aniline in β-cyclodextrin solution with DMAD and phenacyl bromide. The scope and generality of the reaction have been investigated by including various aromatic amines and phenacyl bromides. The recyclability of β-cyclodextrin was established in this environmentally benign protocol.

Synthesis of N-substituted pyrroles was described by Fu-Jun Duan et al. [20], and the title compounds were obtained by Paal-Knorr reaction of γ-diketones with amines in presence β-cyclodextrin in aqueous medium

(Figure 15). The scope and limitations of this approach were examined and explored to include a wide variety of aryl, aralkyl, and alkylamines. It was reported that the role of β-cyclodextrin was to facilitate the formation of N-substituted pyrroles via the inclusion of γ-diketones stabilized by the hydroxyl groups of cyclodextrin, subsequent cyclization followed by elimination leading to the products.

A facile regioselective synthesis of polysubstituted pyrroles was presented by Rajendran Suresh et al. [21] from 1,3-dicarbonyl compounds and α-azido chalcones catalyzed by indium trichloride in water medium (Figure 16).

α-Azido chalcones were prepared from the corresponding acetophenone following previously published papers. Azido chalcones on treatment with acetylacetone in toluene at reflux resulted in polysubstituted pyrroles as end products.

Authors extensively studied the initial reaction and worked on different solvents and solid acid catalysts for enhancing the efficiency and selectivity of the reaction. Excellent results were reported when the reactions were carried out in water catalyzed by indium chloride as lewis acid. The synthetic utility of the reaction was explored further by expanding to several α-azido chalcones and various 1,3-dicarbonyl compounds. The possible mechanism was discussed in the paper.

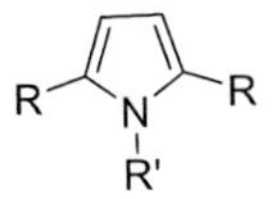

R = CH_3, Ph

R' = C_6H_5; 4-CH_3-C_6H_4; 4-Cl-C_6H_4; 4-F-C_6H_4; 4-CF_3-C_6H_4; 4-OCH_3-C_6H_4; 2,6-$(CH_3)_2$-C_6H_3; C_6H_5-$CH2$; C_6H_5-$CH2$-$CH2$; C_4H_9; 2,6-i-propyl-C_6H_3

Figure 15. Different N-substituted pyrroles.

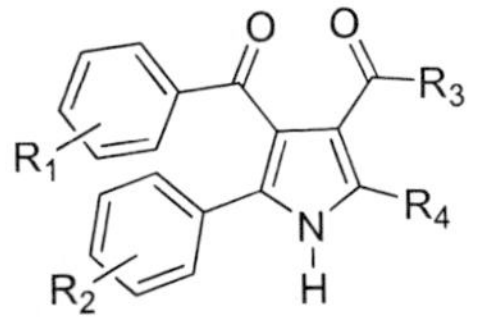

R_1 = R_2 = H, H; H, Cl; CH_3, Cl; Cl, H; Cl, CH_3; Cl, Br; Br, Cl; Br, NO_2; CH_3, H; OCH_3, Cl; NO_2, OCH_3

R_3, R_4 = CH_3, CH_3; CH_3, CF_3; CH_3, Ph

Figure 16. Various polysubstituted pyrroles.

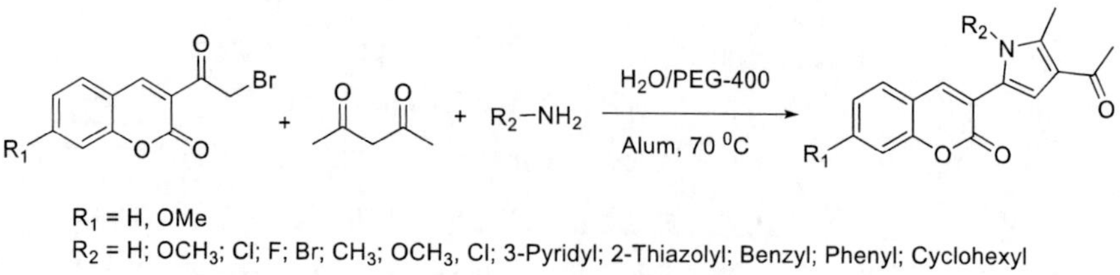
Br
+
+ R2−NH2
H2O/PEG-400
Alum, 70 °C
R1
R2
N
O
R1 = H, OMe
R2 = H; OCH3; Cl; F; Br; CH3; OCH3, Cl; 3-Pyridyl; 2-Thiazolyl; Benzyl; Phenyl; Cyclohexyl

Figure 17. Synthesis of 3-(1H-pyrrol-2-yl)-2H-chromen-2-ones.

Gargi Pal et al. [22] in their interesting article explained the synthesis of 3-(1H-pyrrol-2-yl)-2H-chromen-2-ones by a one-pot three-component approach catalyzed by inexpensive, eco-friendly and nontoxic alum in water-PEG binary solvent medium from 3-(bromoacetyl)-coumarin, acetylacetone and amine (Figure 17). Authors claim shorter reaction times, simplified procedure, excellent yields, and avoiding any toxic catalysts, solvents as advantages of the present protocol. Various catalyst systems were screened and compared for this reaction in the optimization studies and found that alum was proved to be the most efficient catalyst. Best results were reported in water-PEG 400 binary system compared to others such as THF, MeCN, MeOH, and toluene.

The authors explained a plausible mechanism for the formation of pyrrole derivatives. It was explained that using a binary system as a medium enhanced not only the solubility of the reactants but also the catalytic activity of alum, due to the coordination of the hydroxyl group of PEG-400 with Al^{+3} species of the catalyst, as evidenced by FT-IR spectra of the alum (PEG-Al) inclusion complex. Single-crystal X-ray studies supported the structure of the product. The methodology was extended to synthesize a wide range of novel pyrrole derivatives.

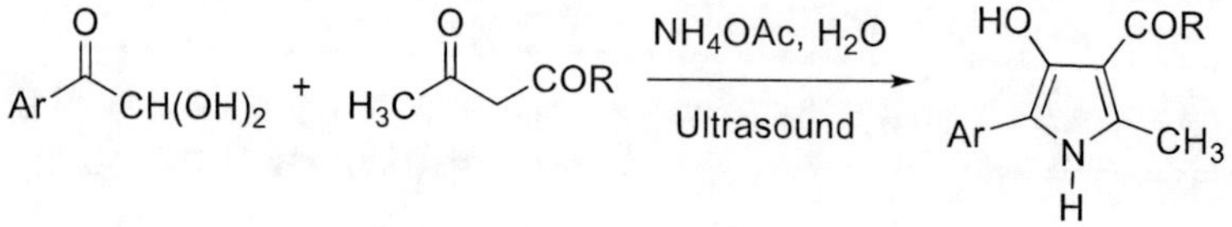
O
Ar
CH(OH)2
+
H3C
COR
NH4OAc, H2O
Ultrasound
HO
COR
Ar
N
H
CH3

Figure 18. Synthesis of 5-aryl-4-hydroxy-2-methyl-1H-pyrrole-3-carboxylic acid esters.

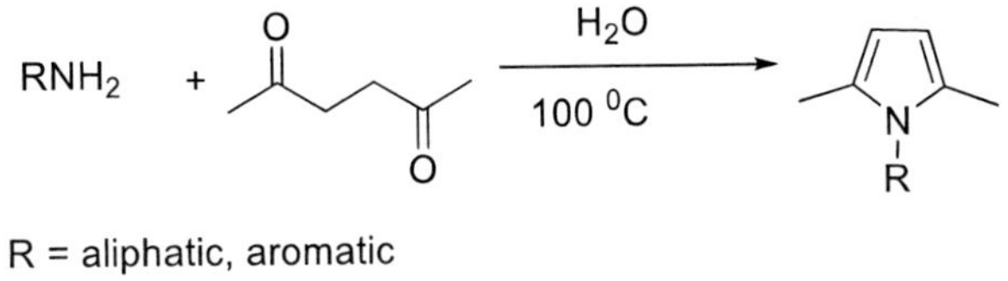

RNH$_2$ + [β-diketone] $\xrightarrow[100\ ^{\circ}C]{H_2O}$ [pyrrole]

R = aliphatic, aromatic

Figure 19. Synthesis of 1-benzyl-2,5-dimethyl-1H-pyrrole derivatives.

Bagher Eftekhari-sis and SalehVahdati-Khajeh [23] reported a green and efficient ultrasound-assisted synthesis of 5-aryl-4-hydroxy-2-methyl-1H-pyrrole-3-carboxylic acid esters and 6-aryl-3-methylpyridazine-4-carboxylic acid esters *via* a three-component reaction of arylglyoxal hydrates with β-dicarbonyl compounds in the presence of hydrazine hydrate and ammonium acetate in aqueous medium (Figure 18). As proposed by the authors, the reaction mechanism involved the attack of enamino ester produced in situ onto the glyoxal to obtain another enamino ester intermediate. The nucleophilic addition of amine group of this intermediate on to the second carbonyl group of phenylglyoxal part followed by dehydration afforded the hydroxyl pyrrole derivatives.

Various N-substituted pyrroles were prepared in aqueous phase by the reaction of diversified aliphatic and aromatic primary amines with hexa-2,5-dione in a simple and eco-friendly approach as described by Dilek Akbaslar et al. [24] Initial studies were carried out on the synthesis of 1-benzyl-2,5-dimethyl-1H-pyrrole as a model compound in water at different temperatures (Figure 19). The process was extended to include different aliphatic and aromatic primary amines after establishing the optimum conditions. Furthermore, it was observed that aliphatic amines provided better yields than aromatic counterparts. Bipyrrole and tripyrrole derivatives were obtained from di and tri amino substrates, respectively. Extensive studies on the preparation of some more selected pyrrole analogues in different organic solvents as well as in water suggested that the yields were almost similar, confirming the role of water as an environmentally benign solvent. Interestingly, the authors also presented a comparative study of the present protocol with different methodologies utilizing various catalysts and solvents, to conclude the efficacy of the present approach, citing the preparation of 2,5-dimethyl-1-phenyl-1H-pyrrole as a model compound.

Hojat Veisi et al. [25] in an interesting research paper described the preparation of sulfamic acid-functionalized magnetic Fe_3O_4 nanoparticles (MNPs/DAG-SO_3H) as an active, recyclable and stable magnetically separable acidic nanocatalyst and applied to the one-pot synthesis of N-substituted pyrroles from primary amines and γ-diketones in aqueous phase by Paal-Knorr condensation at room temperature (Figure 20). Immobilization of sulfamic acid groups on amino-functionalized Fe_3O_4 nanoparticles (Fe_3O_4-Diaminoglyoxime) afforded the active catalyst, which was characterized adequately by XRD, FT-IR, and TEM before utilization. Authors mention that such inorganic-organic hybrids have an advantage of the favorable combination of both organic and inorganic properties in single nanomaterial.

Systematic studies were carried out to determine the catalyst loading as well as to select the appropriate solvent. Attractive advantages of this methodology are the rapid and easy separation of the catalyst employing a suitable external magnet, which minimizes catalyst loss during the separation process.

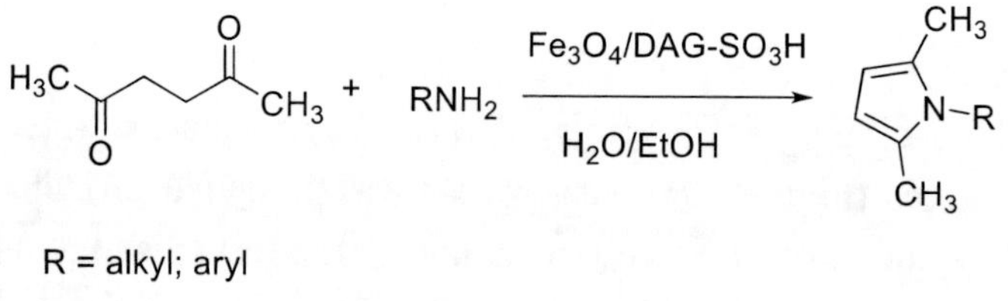

Figure 20. Synthesis of N-substituted pyrroles.

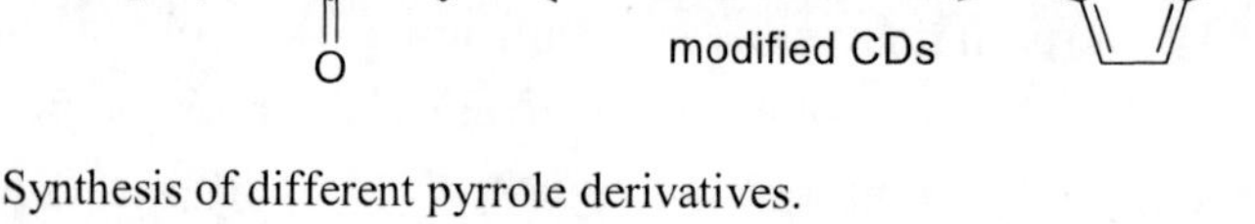

Figure 21. Synthesis of different pyrrole derivatives.

Stephane Menuel et al. [26] developed a new and efficient synthetic protocol to access pyrrole derivatives from amines and 1,4-diketones by Paal-Knorr condensation strategy in aqueous phase using partially methylated cyclodextrins (RAME-β-CD) as mass transfer agents (Figure 21). Cyclodextrins are water-soluble, ring-shaped oligomers with truncated cone-shaped structures. CDs have been selectively modified to change their properties. Even though some of these are commercially available, others are being synthesized and studied for their modified properties. Among these randomly methylated cyclodextrins (RAME-CD) are more soluble, surface-active as well as reported [27] to be more efficient as mass transfer promoters for catalytic reaction in biphasic media.

Authors reported Paal-Knorr reaction with different modified cyclodextrins as mass transfer agents under aqueous biphasic conditions. The reaction of aniline and 2,5-hexanedione was initially investigated under different reaction conditions. It was observed that in the presence of RAME-β-CD, the rate of N-substituted pyrrole formation was significant. Authors postulate that the interaction between aniline and non-methylated-OH of cyclodextrin can occur during the inclusion process and activate the pyrrole formation. The formation of inclusion complex between aniline and RAME-β-CD was confirmed by TROESY experiments. It was observed that RAME-β-CD provided better results than native analogues. After optimization, the reaction conditions were extended to include a wide range of aromatic and aliphatic amines. It was interesting to note that selective mono functionalization of diamines was also achieved by the authors in this study, utilizing RAME-β-CD as mass transfer agent in the absence of an acid catalyst. The scope of the approach was also examined with more hindered and less reactive ketones, and improved yields were observed with p-nitrophenol as a weak acid catalyst. Authors also claim that the present methodology has the advantages of performing the reaction under nearly neutral and mild conditions to allow pyrrole formation with functionalized groups such as amines or alcohols and also direct access to unsymmetrical bispyrroles.

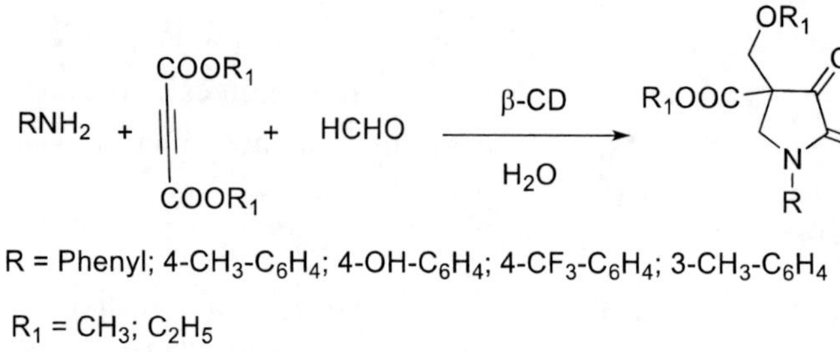

RNH$_2$ + (CH(COOR$_1$)≡C(COOR$_1$)) + HCHO $\xrightarrow[\text{H}_2\text{O}]{\beta\text{-CD}}$ [pyrrolidine product with R$_1$OOC, CH$_2$OR$_1$, =O, N-R]

R = Phenyl; 4-CH$_3$-C$_6$H$_4$; 4-OH-C$_6$H$_4$; 4-CF$_3$-C$_6$H$_4$; 3-CH$_3$-C$_6$H$_4$

R$_1$ = CH$_3$; C$_2$H$_5$

Figure 22. Synthesis of highly substituted pyrrolidine derivatives.

J. Shankar et al. [28] reported the aqueous phase synthesis of highly substituted pyrimidine/pyrrolidine derivatives under neutral conditions by the reaction of aromatic amines, dimethyl/diethyl acetylenedicarboxylate and formaldehyde mediated by inexpensive, readily available, non-toxic and recyclable β-cyclodextrin (Figure 22).

The reactions were carried out by the in situ formation of a β-CD complex of amine in water followed by the addition of but-2-ynedioate and formaldehyde. NMR studies confirmed the complex formation between aniline and β-cyclodextrin. The complexation increases the reactivity of the aniline group of the molecule due to the intermolecular hydrogen bonding with CD-hydroxyl groups, which facilitate the addition of but-2-ynedioates. Furthermore, the catalytic efficiency of the recovered β-CD was checked.

A wide variety of 3-substituted 2-aryl-1H-pyrroles were prepared by Pd(II) catalyzed C-C coupling reaction between substituted aliphatic nitriles and diversely substituted aryl boronic acids in a convenient one-pot eco-friendly and atom economical approach as described by Md Yousuf and Susanta Adhikari [29].

This approach involves the addition of an aryl group of aryl boronic acid to substituted nitriles followed by amination-annulation-dehydration sequence. During optimization studies, authors screened different Pd sources, additives as well as solvents. Among various Pd sources, Pd(OAc)$_2$ was observed to be excellent for catalytic efficiency. It was reported that water-acetic acid mixture as medium gave better yields. The generality of the protocol was examined to include various electronically and structurally diverse aryl boronic acids (Figure 23). The authors observed the lack of reactivity in case of electron-withdrawing groups on aryl boronic acids due to the lower nucleophilicity of the aryl groups of aryl boronic acids. Moreover, sterically hindered aryl boronic acids also were observed to afford good amounts of the products in this interesting protocol.

Scope of the reaction of various boronic acids was extended to both substituted ethyl cyanoacetate as well substituted malononitriles. Structures of the final representative products were examined by using single-crystal X-ray diffraction studies. Mechanism of the reaction was also explained in this paper. It was claimed that the broad scope of substrates, functional group tolerance, and higher product yields were the advantages of this approach.

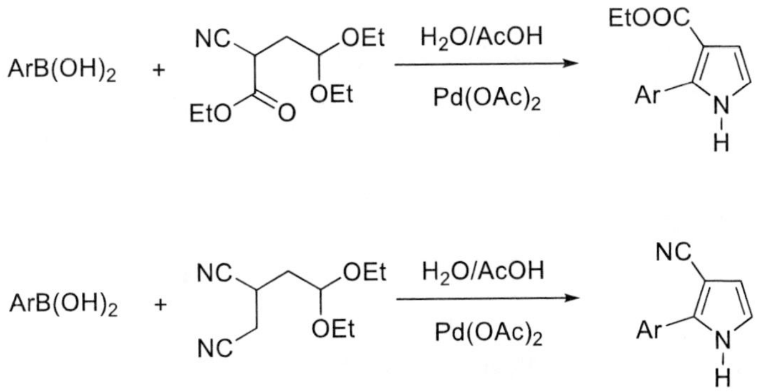

Figure 23. Synthesis of 3-substituted 2-aryl-1H-pyrroles.

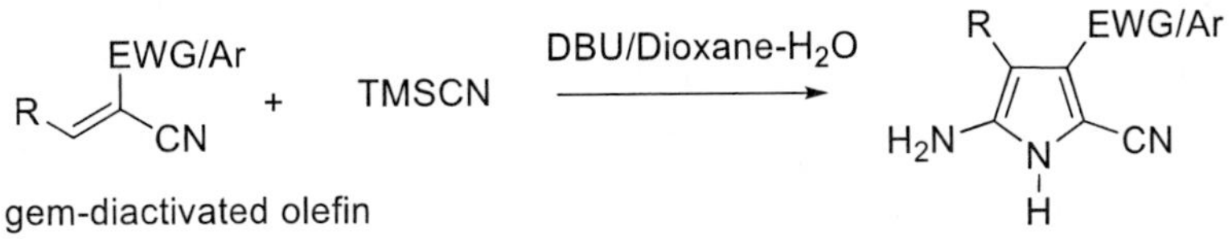

Figure 24. Synthesis of tetra substituted 2-amino-N-H-pyrroles.

A new approach for the preparation of tetra substituted-N-H-pyrroles from gem-diactivated acrylonitriles and TMSCN was developed *via* a tandem process of Michael addition and intramolecular cyanide-mediated nitrile to nitrile condensation by Sankar. K. Guchhait et al. [30]. This reaction constructs multiple C-C/C-N bonds and also involves C/N nucleophilic and electrophilic roles of nitrile. Dioxane-water mixture was reported to be effective, and water was reported to be acting as a proton source and to induce H-bond driven activation of functionalities (Figure 24). During the investigation, the efficiency of several commonly used metal and nonmetallic cyanating agents were studied. It was observed that TMSCN not only acted as a cyanide source but also played an additional role. It was suggested that the silyl byproduct generated from TMSCN acted as a lewis acid in electrophilic activation of the substrate as well as the intermediate. TMSCN was expected to provide multiple C-C/C-N bond constructions and produce poly functionalized pyrroles. The new methodology was utilized to prepare versatile tetra substituted-2-amino-N-H-pyrroles. TMSCN underwent reaction with alkylidene malononitriles possessing different aryl, heteroaryl, alkyl and other functionalities. It was reported that aryl groups with various electronic and steric properties were compatible, though electron-withdrawing groups gave relatively lesser yields compared to electron-donating functionalities.

A versatile environmentally benign synthesis of pyrroles and N-substituted pyrroles was developed by Rani. N. Patil and A. Vijay Kumar [31] using sulfonic acid functionalized β-cyclodextrin as a supramolecular catalyst under aqueous conditions (Figure 25). CD was found to play a dual role as an acid catalyst as well as phase transferring reagent. Initially, the catalyst was applied for Clauson-Kaas pyrrole synthesis followed by an extension to Paal-Knorr reaction and proved that the acidic functionality of CD plays an active role in catalyzing the reactions. The scope of the catalyst was well explored to include a wide range of aromatic, hetero-aromatic and poly-aromatic amines to afford pyrrole derivatives in excellent yields. Recyclability and catalyst loading evaluations were reported. To gain insights into the involvement of β-CD catalyst into the reaction mechanism, authors employed in situ NMR studies and confirmed the formation of aniline-cyclodextrin inclusion complex.

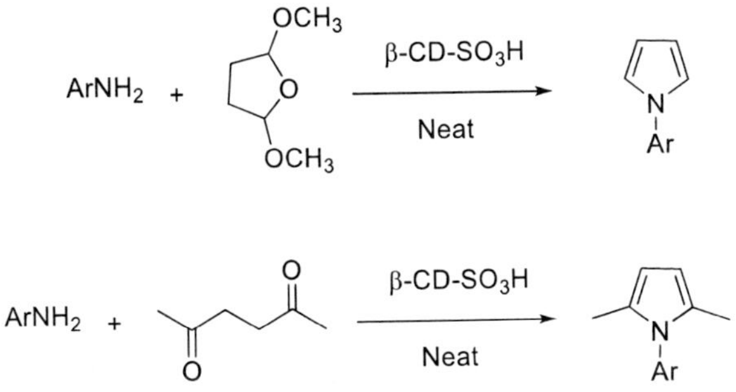

Figure 25. Synthesis of N-substituted pyrrole derivatives.

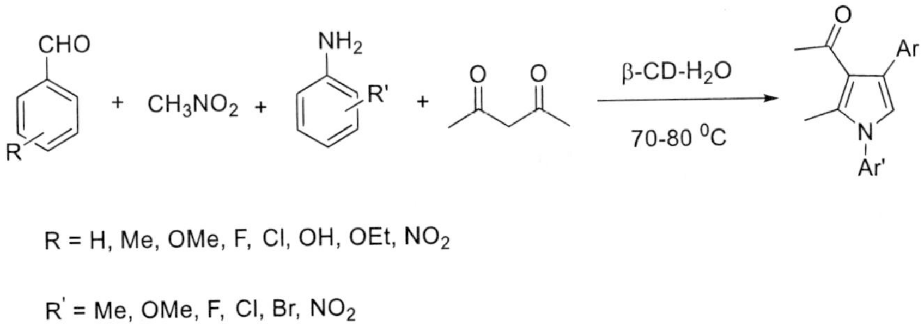

R = H, Me, OMe, F, Cl, OH, OEt, NO_2

R' = Me, OMe, F, Cl, Br, NO_2

Figure 26. Synthesis of poly functionalized pyrroles.

Karnakar K. et al. [32] reported an elegant green approach for the synthesis of poly functionalized pyrroles in the presence of β-cyclodextrin as a recyclable supramolecular catalyst in aqueous phase (Figure 26). Cyclodextrins are cyclic oligosaccharides possessing hydrophobic cavities and exhibit host-guest complexation phenomenon. Noncovalent bond interaction between the hydrophobic cavity of the CD and hydrophobic guest molecule leads to inclusion complex formation.

Authors investigated the efficiency of different supramolecular catalysts such as α, β, and γ-cyclodextrins as well as 2-hydroxypropyl-β-cyclodextrin and methyl-β-cyclodextrin and selected β-cyclodextrin as a better choice. The progress of this one-pot four-component reaction was also examined in the presence of sodium dodecyl sulphate, PEG-400, and water. Scope of the reaction was extended to include various aromatic aldehydes and aromatic amines. It was claimed that this protocol was an inexpensive, and useful addition to synthetic organic chemistry to access poly functionalized pyrroles.

CONCLUSION

The last few years have witnessed considerable growth in organic chemistry research towards achieving environmentally benign approaches for synthesizing various heterocyclic systems. Further design and development of greener protocols for the preparation of pyrrole derivatives will ensure the rapid growth of an active and significant area for constructing biologically active pyrrole scaffolds. This academic review is an exercise to bring together the significant contributions of pyrrole synthesis by eco-friendly routes at one place. Scholars are advised to go through original research works for detailed information. The structures are also drawn briefly to give an idea about the products.

REFERENCES

[1] Tukuda, F., Ishibashi, F., Iwao, M. *Heterocycles,* 2011, 83, 491-529.
[2] Williamson, N. R., Simonsen, H. T., Ahmed, R. A., Goldet, G., Slater, H., Woodley, L., Leeper, F. J., Salmond, G. P. *Mol. Microbiol.* 2005, 56, 971-989.
[3] Williamson, N. R., Fineran, P. C., Gristwood, T., Chawrai, S. R., Leeper, F. J., Salmond, G. P. C. *Future Microbiol.* 2007, 2, 605-618.
[4] Boger, D. L., Boyce, C. W., Labrili, M. A., Sehon, C. A., Jin, Q. *J. Am. Chem. Soc.* 1999, 121, 54-62, Paolesse, R., Nardis, S., Sagone, F., Khoury, R. G. *J. Org. Chem.* 2001, 66, 550-556.
[5] Domingo, V. M., Aleman, C., Brillas, E., Julia, L. *J. Org. Chem.* 2001, 66, 4058-4061.

[6] Kaiser, D. G., Glenn, E. M. *J. Pharm. Sci.* 1972, 61, 1908-1911, Daisone, G., Maggio, B., Schillaci, D. *Pharmazie* 1990, 45, 441-442, Almerico, A. M., Diana, P., Barraja, P., Dattolo, G., Mingoia, F., Putzolu, M., Perra, G., Milia, C., Musiu, C., Marongiu, M. E. *Farmaco* 1997, 52, 667-672, Lehuede, J., Fauconneau, B., Barrier, L., Ourakow, M., Piriou, A., Vierfond, J. M. *Eur. J. Med. Chem.* 1999, 34, 991-996, Chong, P. H., Bachenheimer, B. S. *Drugs* 2000, 11, 351-360, Pinna, G. A., Loriga, G., Murineddu, G., Grella, G., Mura, M., Vargiu, L., Murgioni, C., La Colla, P. *Chem. Pharm. Bull.* 2001, 49, 1406-1411, Iwao, M., Takeuchi, T., Fujikawa, N., Fukuda, T., Ishibashi, F. *Tetrahedron Lett.* 2003, 44, 4443-4446.

[7] Yavari, I., Kowsari, E. *Synlett.* 2008, 6, 897-899.

[8] Khalili, B., Jajarmi, P., Eftekhari-Sis, B., Hashemi, M. M. *J. Org. Chem.* 2008, 73, 2090-2095.

[9] Yoshida, M., Al-Amin, M., Shishido, K. *Synthesis.* 2009, 14, 2454-2466.

[10] Yavari, I., Kowsari, E. *Mol. Divers.* 2009, 13, 519-528.

[11] Murthy, S. N., Madhav, B., Kumar, A. V., Rao, K. M., Nageswar, Y. V. D. *Helv. Chimi. Acta.* 2009, 92, 2118-2124.

[12] Miles, K. C., Mays, S. M., Southerland, B. K., Auvil, T. J., Ketcha, D. M. *Arkivoc,* 2009, 14, 181-190.

[13] Azizi, N., Amiri, A. K., Ghafuri, H., Bolourtchain, M., Saidi, M. R. *Synlett.* 2009, 14, 2245-2248.

[14] Rueping, M., Parra, A. *Org. Lett.* 2010, 12, 5281-5283.

[15] Kolontsova, A. N., Ivantsova, M. N., Tokareva, M. I., Mironov, M. A. *Mol. Divers.* 2010, 14, 543-550.

[16] Murthy, S. N., Nageswar, Y. V. D. *Tetrahedron Letters.* 2011, 52, 4481-4484.

[17] Meshram, H. M., Bangade, V. M., Reddy, B. C., Kumar, G. S., Thakur, P. B. *Int. J. Org. Chem.* 2012, 2, 159-165.

[18] Azizi, N., Davoudpour, A., Eskandari, F., Batebi, E. *Montasch Chem.* 2013, 144, 405-409.

[19] Ramesh, K., Karnakar, K., Satish, G., Nageswar, Y. V. D. *Chin. Chem. Lett.* 2012, 23, 1331-1334.

[20] Duan, F. J., Ding, J. C., Deng, H. J., Chen, D. B., Chen, J. X., Liu, M. C., Wu, H. Y. *Chin. Chem. Lett.* 2013, 24, 793-796.
[21] Suresh, R., Muthusubramanian, S., Nagaraj, M., Manickam, G. *Tetrahedron Lett.* 2013, 54, 1779-1784.
[22] Pal, G., Paul, S., Das, A. R. *Synthesis,* 2013, 45, 1191-1200.
[23] Eftekhari-Sis, B., Khajeh, S. V. *Curr. Chem. Lett.* 2013, 2, 85-92.
[24] Akkbaslar, D., Demirkol, O., Giray, S. *Synth. Commun.* 2014, 44, 1323-1332.
[25] Veisi, H., Mohammadi, P., Gholami, J. *Appl. OrganoMetal. Chem.* 2014, 28, 868-873.
[26] Menuel, S., Rousseau, J., Rousseau, C., Vaiciunaite, E., Dodonova, J., Tumkevicius, S., Monflier, E. *Eur. J. Org. Chem.* 2014, 4356-4361.
[27] Hapiot, F., Ponchel, A., Tilloy, S., Monflier, E. *C. R. Chim.* 2011, 14, 149, Bricout, H., Hapiot, F., Ponchel, A., Tilloy, S., Monflier, E. *Curr. Org. Chem.* 2010, 14, 1296.
[28] Shanker, J., Satish, G., Ramesh, K., Nageswar, Y. V. D. *Eur. J. Chem.* 2014, 5, 541-544.
[29] Yousuf, Md., Adhikari, S. *Org. Lett.* 2017, 19, 2214-2217.
[30] Guchhait, S. K., Sisodiya, S., Saini, M., Shah, Y. V., Kumar, G., Daniel, D. P., Hura, N., Chaudary, V. *J. Org. Chem.* 2018, 83, 5807-5815.
[31] Patil, R. N., Kumar, A. V. *ChemistrySelect,* 2018, 3, 9812-9818.
[32] Konkala, K., Chowrasia, R., Manjari, P. S., Domingues, N. L. C., Katla, R. *RSC. Adv.* 2016, 6, 43339-43344.

ABOUT THE AUTHOR

Dr. Y. V. D. Nageswar obtained BSC and MSC degrees from Osmania University, Hyderabad (India) and PhD degree from Kakatiya University, Warangal (India, 1980). Later he joined Indian Institute of Chemical Technology, Hyderabad and elevated to the position of Chief Scientist. He worked as Dean (Academic& Faculty), National Institute of pharmaceutical Education and Research, Hyderabad. Being a synthetic organic chemist, he specialized in green chemistry, supramolecular chemistry, medicinal chemistry and nano catalysis, besides technology development work on several active molecules. He is recognized as a research supervisor by different Indian universities and has published original research articles in highly rated international science journals. He is also a reviewer for various International journals and is a recipient of Fellow of Royal Society of Chemistry (London). He has been an expert member on several state and university boards. He is acting as honorary academic and research advisor, Federal University of Grande Dourados, Brazil.

In: Pyrrole: Synthesis and Applications
Editor: Colin Welch

ISBN: 978-1-53617-137-2
© 2020 Nova Science Publishers, Inc.

Chapter 3

UV AND VISIBLE LIGHT PHOTOINDUCED POLYMERIZATION OF PYRROLE/METHACRYLATE

Claudia I. Vallo and Silvana V. Asmussen[*]

Institute of Materials Science and Technology (INTEMA),
University of Mar del Plata, CONICET, Mar del Plata, Argentina

ABSTRACT

Electrically-conducting polymers such as polypyrrole have been the focus of intense research interest over the last ten decades because they can be used in a wide range of technologies. However, it is well established that electrically-conducting polymers fabricated by chemical or electrochemical polymerization processes are either powders or intractable polymers which exhibit deficient mechanical properties and are difficult to process. A possible way of improving the mechanical properties of polypyrrole is by mixing it with other polymers in order to reach a synergetic overall performance. In this study pyrrole was blended with a methacrylate resin and the mixtures were processed by photopolymerization. This polymer processing method has the advantage

---

[*] Corresponding Author's Email:civallo@fi.mdp.edu.ar.

that permits the incorporation of different additives and flexibilizers into the resins thereby optimizing its manufacture and improving the mechanical properties of the final cured material. The photoinitiator systems used to cure the mixtures pyrrole/methacrylate consisted of the iodonium salt p-(Octyloxyphenyl)phenyliodonium hexafluoroantimonate (IODS), in combination with Benzil α,α-dimethyl acetal (BDMA), α-Methoxy-α-phenylacetophenone (MPAP) or the pair camphorquinone (CQ)/ethyl-4-dimethylamino benzoate (EDMB). Mixtures photoactivated with the IODS salt in combination with BDMA or MPAP were efficiently cured under UV irradiation (λ=365 nm). On the other hand, in mixtures photoactivated with IODS/CQ/EDMB and irradiated with visible light ((λ=470 nm) the polymerization of both methacrylate and pyrrole was much slower. Scanning electron microscopy studies showed no sign of phase separation demonstrating that the pyrrole/methacrylate blends formed an interpenetrating network. Studies of electrical properties of the hybrid polymers revealed that their electrical conductivity increased markedly with the proportion of pyrrole in the initial mixture. This is attributed to the formation of an electrically conducting polymer network in the non-conducting methacrylate matrix.

Keywords: pyrrole, photopolymerization, methacrylate, iodonium salt, condudting polymers

INTRODUCTION

Polypyrrole and derivatives are conducting polymers which have attracted great interest over the last two decades for many applications such as corrosion resistant coatings [1], the preparation of rechargeable batteries [2], printed circuit boards manufacture [3], conducting inks [4] and antibacterial silver nanoparticles coated with polypyrrole [5].

π-conjugated electronically conducting polymers, such as polypyrrole, are mostly synthesized by either chemical [6-9] or electrochemical processes [10-12]. Chemical synthesis is carried out by the oxidation of pyrrole with a proper oxidant such as ferric chloride while electrochemical oxidation of pyrrole results in a layer of electronically conducting polypyrrole at the surface of the electrode [6]. Unfortunately, these methods produce an intractable solid polymer or powder which displays low processability and poor mechanical properties. Many efforts have been carried out in order to improve the overall properties of polypyrrole by forming blends with other polymers. For example, the polymerization of pyrrole has been carried out inside porous poly(vinyl alcohol) [13], polymethylmethacrylate [14], and poly(vinylchloride) matrixes [15]. Electrical conductive films based on cellulose acetate- pyrrole have been recently, prepared by Takano et al. [16]. In another study, Migahed et al. [17] reported the synthesis of polypyrrole in ethylene-vinylalcohol copolymer.

Although extensive results have been reported on electrochemical and chemical polymerization of pyrrole, studies on the photoinduced polymerization are relatively limited [18-23]. The photopolymerization process yields significant advantages because different comonomers, additives and flexibilizers can be incorporated into light-curing formulations in order to improve the mechanical properties of the resultant polymers. The objective of our research was to study the photoinduced polymerization of blends of pyrrole and a bifunctional methacrylate monomer. Polymers prepared from mixtures pyrrole/methacrylate have the advantage of combining the electronic conductivity of polypyrrole with the simpler processing techniques and attractive mechanical characteristics of methacrylate-based polymers. To the best of our knowledge, no studies concerning the photopolymerization of pyrrole and methacrylate resins have been previously reported. Pyrrole polymerizes by a cationic mechanism while methacrylate resins are readily photocured by a free radical mechanism. The pyrrole/methacrylate blend was photoactivated with an iodonium salt in combination with either UV of visible photoinitiator systems. Benzil α,α-dimethyl acetal and α-methoxy-α-phenylacetophenone were used as UV photoinitiators while the pair camphorquinone/ethyl-4-dimethylamino benzoate was used as visible light photoinitiator. The degree of conversion of the individual monomers was monitored by UV-vis spectroscopy and Fourier transform infrared in the mid region (MIR) [24-25]. Conversion values at the surface of thick specimens (~ 2 mm) were evaluated by attenuated total reflectance (ATR).

OUR RESEARCH

This chapter is organized as follows. First, we show the photopolymerization process of mixtures pyrrole/methacrylate monitored by UV-vis spectroscopy. Then, a description of the degree of conversion of the individual monomers by FTIR is presented. Finally, our results on electrical conductivity characterizations are described.

Polymerization of Pyrrole Assessed by UV-vis Spectroscopy

In this section, we describe the polymerization of pyrrole in mixtures pyrrole/methacrylate (Py/MA) photoactivated with both radical and cationic photoinitiators.

Pyrrole (Py) (Sigma Aldrich, Argentina) was double distilled under reduced pressure and then stored at around 5 °C. The methacrylate monomer 2,2- bis[4-(2-methacryloxyethoxy)phenyl]propane (MA) was from Esstech, Essington, USA. The iodonium salt was p-(octyloxyphenyl) phenyliodonium hexafluoroantimonate (IODS) (OMAN 071, Gelest Inc., Philadelphia, Pennsylvania, USA). The free radical photoinitiators benzil α,α-dimethyl acetal (BDMA), α-methoxy-α-phenylacetophenone (MPAP), Camphorquinone (CQ) and Ethyl-4-dimethyl aminobenzoate (EDMB) were from Sigma Aldrich, Argentina. All materials were used as received without further purification. The chemical structure of the resins and photoinitiators is shown in Figure 1. Two different light sources assembled from Light Emitting Diodes (LEDs) were used in the present study. These LEDs were selected because they resulted in an optimal overlap between their spectral irradiances and the molar absorption spectra of the BDMA, MPAP and CQ. Mixtures Py/MA containing BDMA and MPAP were irradiated with a LED with its emmitance centered at 365 nm. The irradiance of this 365 nm LED was set at 75, 135 and 175 mW/cm^2 by varying the electrical voltage through the semiconductor: The mixtures pyrrole/MA photoactivated with CQ/EDMB were irradiated with a LED having a wavelength range 410–530 nm, peak at 470 nm and irradiance equal to 600 mW cm^{-2} (Valo, Ultradent, USA).

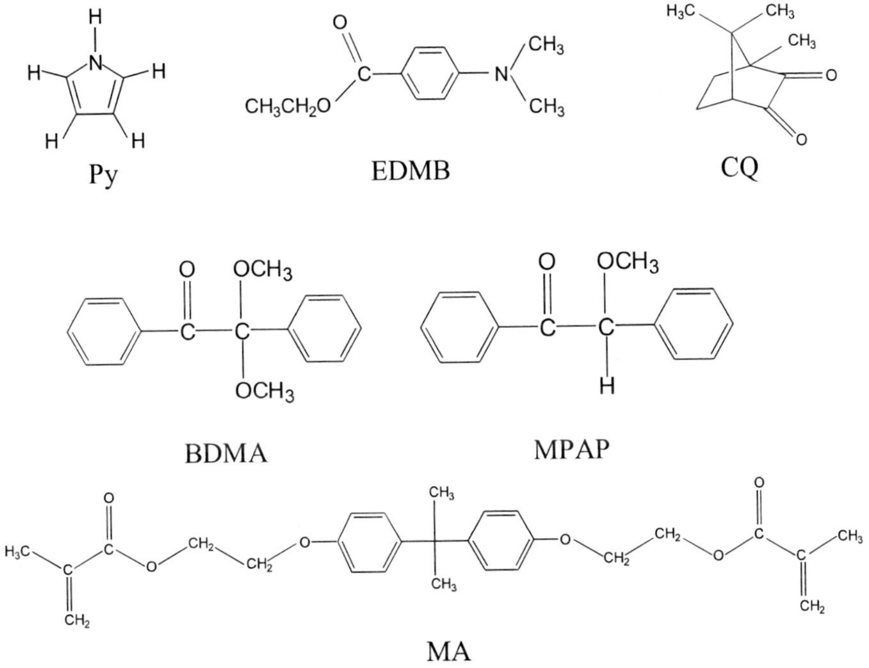

Figure 1. Chemical structure of the photoinitiator and monomers studied.

The UV-vis absorption spectra of mixtures Py/MA were acquired using a UV-visible spectrophotometer (1601 PC, Shimadzu) at room temperature (*ca* 20 ±2°C). UV-vis studies in an air environment were carried out in 0.5 ± 0.1 mm thick samples sandwiched between two disposable 1 mm thick quartz plates. The appearance of polypyrrole was monitored using the changes in absorbance at the wavelength of its maximum absorption. The molar absorption coefficients of MPAP and BDMA in MA at 365 nm are 110 (l/mol cm) and 167 (l/mol cm) respectively while the molar absorption coefficient of CQ in MA at 467 nm is 42 (l/mol cm).

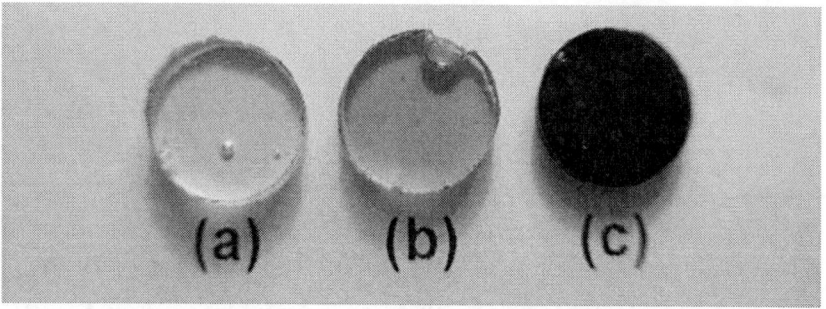

Figure 2. Photopolymerized samples of MA/Py mixtures (a) MA/Py 97:3 by weight in the absence of IODS (b) MA/Py 97:3 by weight containing of 2 wt.. % IODS (c) MA/Py 60:40 by weight containing 2 wt.. % IODS. The intensity of the brown color depends on the proportion of pyrrole in the mixtures.

The mixtures Py/MA were photoactivated with 1 wt.. % of BDMA, MPAP or CQ/EDMB either in the absence or the presence of 2 wt.. % of diaryliodonium salt (IODS). Mixtures containing of BDMA or MPAP were irradiated with the 365-nm LED, while mixtures containing CQ/EDMB were irradiated with the 470 nm LED. Polymerized mixtures MA/Py are shown in Figure 2, which sows that samples photoactivated with BDMA in the absence of IODS were colorless and transparent (Figure 2 (a)) because in the absence of the diaryliodonium salt the polymerization of Py did not occurr. On the other hand, samples prepared with diaryliodonium salt exhibited brown color associated with the formation of PPy. This result clearly demonstrate that, under the experimental conditions used in this study, the methacrylate resin polymerizes by a free radical mechanism photoinitiated by BDMA, MPAP or CQ/EDMB, while pyrrole polymerizes by a cationic mechanism photoinitiated by the diaryliodonium salt. Figure 3 illustrates the proposed mechanism of formation of polypyrrole [9]

The progress of the photoinduced polymerization of pyrrole in mixtures MA/Py 97:3 by weight, containing different photoinitiator systems, was studied by UV-visible spectroscopy. This method measures accurately absorbance values up to 2.5, therefore, only mixtures with low proportion of pyrrole could be analyzed. All samples were irradiated at selected time intervals. Spectra were acquired immediately after each irradiation interval. In addition, the progress of the polymerization reaction was followed in the absence of radiation. Figure 4 a-c show the UV-visible spectra of the polymerization product collected at various time intervals. Py, MA, IODS, BDMA and MPAP do not absorb in the range 400-700 nm. Consequently, the increase in absorbance with time is exclusively attributed to the formation of PPy. It should be mentioned that the polymerization of Py in the presence of BDMA/IODS was so fast that a low irradiance source (75 mW/cm^2) had to be used to assess changes in absorbance occurring in the initial stage of the polymerization. Figure 4-a shows that in the mixture containing the pair MPAP/IODS as the reaction advances two bands develop at 489 and 550 nm. These bands are attributed to the π-π* transition and bipolaron excitations respectively (see Figure 3) [26-27]. The spectra of the mixture containing BDMA/IODS in Figure 4-b exhibit two overlapping bands, the π-π* transition band at 429 nm and the bipolaron band at around 480 nm. The different red-shift values of the UV-vis bands of the mixture photoactivated with MPAP/IODS and BDMA/IODS are associated to a comparatively greater size of the oligomers formed in resins containing MPAP [26]. The influence of the photoinitiator used upon the length of the resulting oligomers will be discussed in detail later. Figure 4-c shows the UV-vis spectra of the Py/MA mixture photoactivated with the pair CQ/EDMB. CQ absorbs in the wavelength range 410–510 nm because of the n,π* transition of the α-dicarbonyl chromophore, which overlaps with the π-π* transition accompanying the formation of PPy. Furthermore, the absorbance increase during formation of PPy is accompanied with a decrease of absorbance due to the photobleaching of CQ under irradiation [28]. Consequently, the polymerization of pyrrole photoactivated with the CQ/EDMB pair under irradiation at 467 nm could not be monitored by UV-vis spectroscopy. Alternatively, the rate of polymerization of Py could be assessed from UV-vis spectra after the light source was switched off. Changes in absorbance in the dark, i.e., in the absence of photobleaching of the CQ, are solely attributed to the formation of PPy.

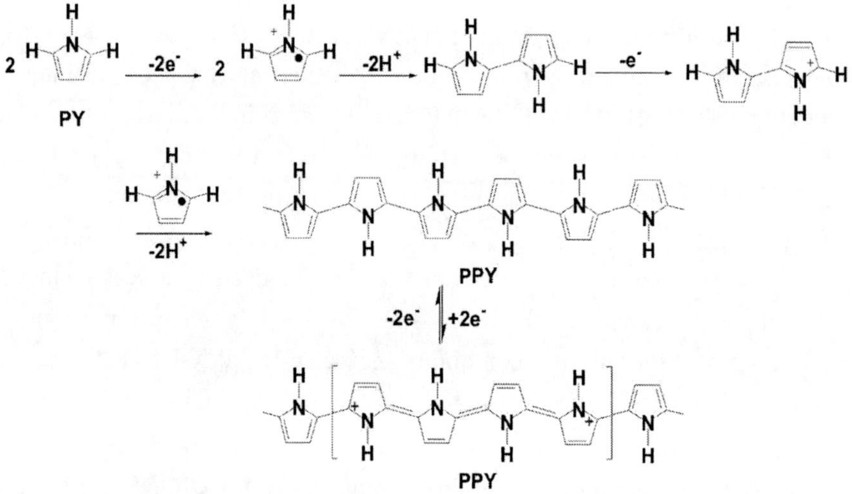

Figure 3. Mechanism of polymerization of pyrrole.

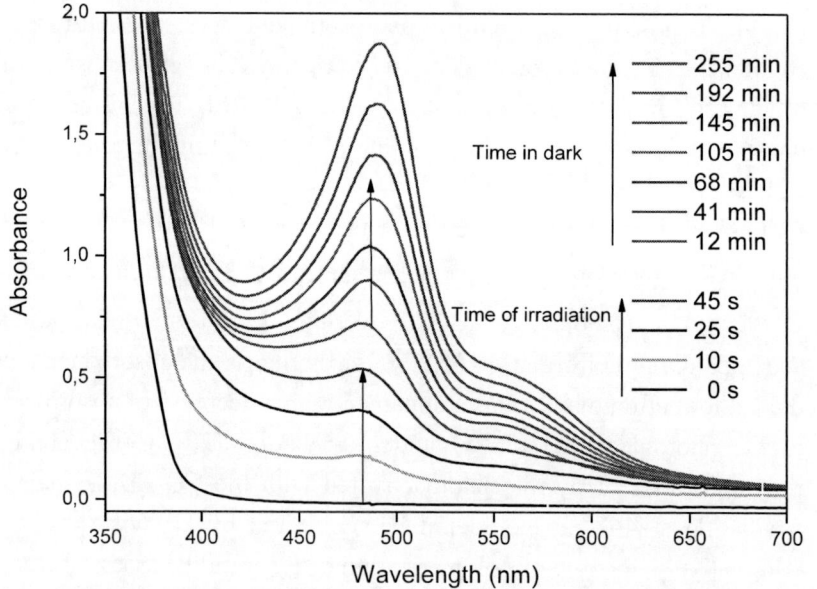

Figure 4(a). UV–vis spectra during polymerization of pyrrole in a mixture MA/Py 97:3 by weight photoactivated with 2 wt.% IODS and 1 wt.% MPAP. Samples were irradiated with the 365 nm LED of 175 mW/cm^2 up to 45 s. After 45 s irradiation the absorbance was monitored in the absence of irradiation.

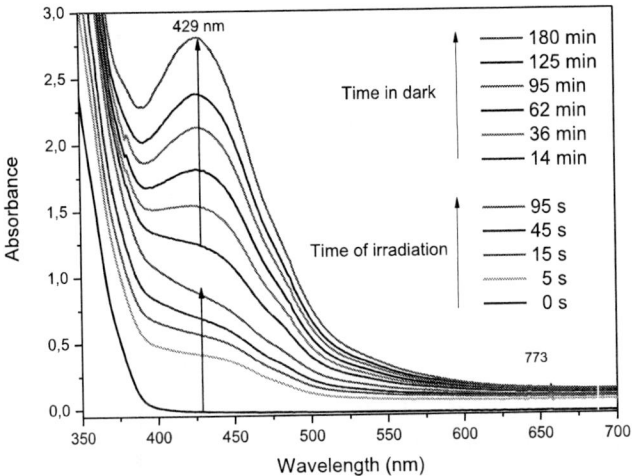

Figure 4(b). UV–vis spectra during polymerization of Py in a mixture MA/Py 97:3 by weight photoactivated with 2 wt.% IODS and 1 wt.% BDMA. Samples were irradiated at regular time intervals up to 95 s with a 365-nm LED of 75 mW/cm^2. After 95 s irradiation spectra were acquired in the dark.

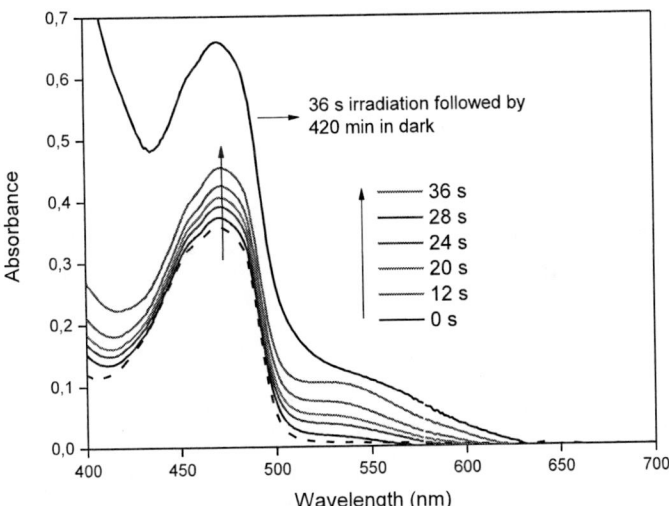

Figure 4 (c). UV–vis spectra during polymerization of Py in a mixture MA/Py 97:3 by weight photoactivated with 2 wt.% IODS and 1 wt.% CQ/EDMB. Samples were irradiated up to 36 s with the 470-nm LED of 600 mW/cm^2. After 36 s rradiation spectra were acquired in the dark. The spectrum at t = 0 corresponds to the absorbance of the CQ.

Figure 5 illustrates the increase in absorbance associated to the π-π* transition band in resins photoactivated with BDMA and MPAP. It shows that pyrrole polymerizes in the absence of light due to the living character of the cationic polymerization mechanism. Results depicted in Figure 5 show that the formation of polypyrrole in mixtures photoactivated with BDMA was significantly faster than that in mixtures containing MPAP. It is worth noting that mixtures containing BDMA/IODS were irradiated with a LED of reduced intensity in order to examine absorbance changes during the early stage of the polymerization. On the other hand, the polymerization of Py in the mixtures photoactivated with CQ/EDMB (not shown in Figure 5) was very slow. Figure 4-c shows that the absorbance value increased from 0.45 to 0.65 after 7 h in the absence of radiation; i.e., in the absence of photobleaching of CQ. The greater efficiency of BDMA compared with MPAP for the polymerization of Pyrrole will be explained along with results of conversion by FTIR spectroscopy.

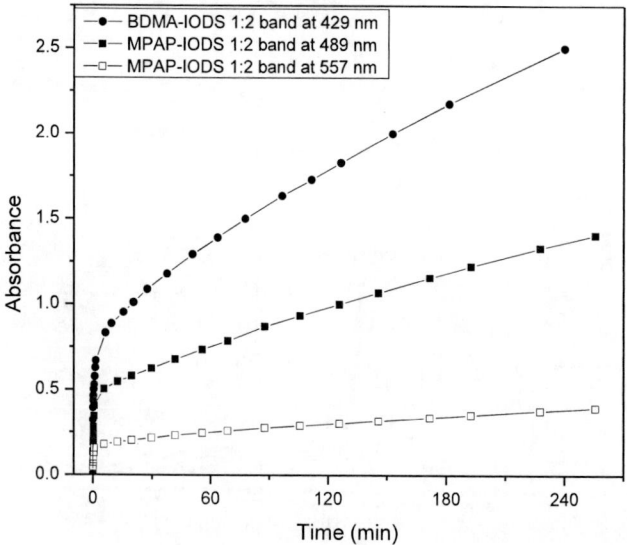

Figure 5. Absorbance associated to the π-π* bands vs time in mixtures MA/Py 97:3 by weight photoactivated with different photoinitiator systems. The absorbance of the polaron excitation resins photoactivated with MPAP is also shown. The irradiance in mixtures with BDMA was 75 mW/cm^2 while in mixtures with MPAP was 175 mW/cm^2.

From results obtained by UV-vis studies it may be concluded that the IODS in combination with BDMA or MPAP are efficient photoinitiator systems under irradiation at 365 nm. On the other hand, in blends photoactivated with IODS/CQ/EDMB the polymerization of both pyrrole and methacrylate groups was comparatively sluggish.

Degree of Conversion of Pyrrole and Methacrylate Monomers Measured by FTIR Spectroscopy

It is worth mentioning that several techniques were explored to measure the conversion of the individual monomers. NIR spectroscopy could not be used to simultaneously monitor the conversion of MA and Py due to the overlap of the bands of MA (6167 cm^{-1}) and Py (6156 and 6113 cm^{-1}) in the region of interest [29, 30]. In addition, the resins exhibited a strong fluorescence during Raman spectroscopy experiments. Consequently, the degree of conversion in mixtures Py/MA containing high molar ratios of Py was assessed by mid-infrared spectroscopy (MIR). MIR spectra were acquired over the range 400-4000 cm^{-1} from 32 co-added scans at 4 cm^{-1} resolution using a Nicolet 6700 Thermo Scientific in transmission mode.

The mixtures Py/MA were sandwiched between two polymer films and were tightly attached to the sample holder using small clamps. The polymer films were selected because of the absence of overlapping of its characteristic bands with those of MA and pyrrole. The background spectra were collected through an empty polymer assembly. With the assembly in a vertical position, the light source was placed in contact with the polymer film surface. The specimens were irradiated at regular time intervals and spectra were collected immediately after each exposure interval. The conversion of methacrylate groups was calculated from the decay of the absorption band located at 1637 cm^{-1} [24]. The conversion of pyrrole was calculated from the decay of the band at 736 cm^{-1} assigned to C-H out-of-plane bending in the pyrrole ring [25]. The band centered at 1510 cm^{-1} was used as internal reference. Figure 6 illustrates typical MIR spectra of a mixture MA/Py. The strong absorption bands at 1074, 1014 and 1049 cm^{-1} are associated to the in-plane deformation vibration of the pyrrole ring [25] while the strong band at 736 cm^{-1} is assigned with the out-of-plane deformation of the pyrrole ring [25]. The conversion of pyrrole was calculated from the decay of the band at 736 cm^{-1} due to the overlap of the bands of pyrrole in the range 1000-1100 cm^{-1} with those of the methacrylate groups. The band at 1637 cm^{-1} (see Figure 6), assigned to the C=C stretching vibrations, was used to calculate the conversion of methacrylate groups [24].

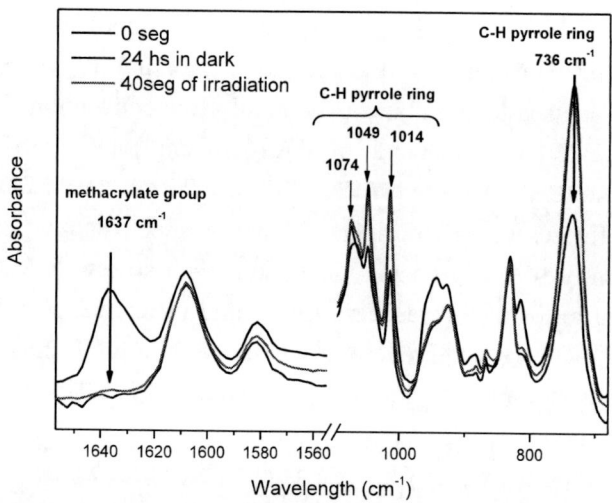

Figure 6. Bands of Py/MA mixtures in MIR spectra. The bands at 1074, 1049, 1014 and 736 cm^{-1} are assigned to Py while the band at 1637 cm^{-1} are from MA.

Typical plots of conversion of Py and MA in mixtures Py/MA 30:70 by weight photoactivated with both cationic and radical type photoinitiators are shown in Figures 7-8.

Plots in Figure 7 demonstrate that the polymerization rate and final extent of conversion of the methacrylate groups are raised by the presence of pyrrole monomer. This trend has also been noticed in mixtures epoxy/methacrylate [30]. The conversion of methacrylate groups after 40 s irradiation was 0.65 in the neat MA photoactivated with BDMA and 0.88 in the mixture Py/MA 30/70 photoactivated with BDMA/IODS.

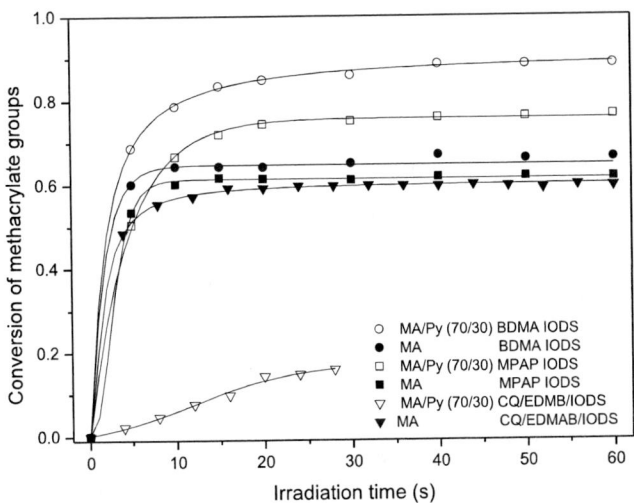

Figure 7. Conversion of C=C groups measured by MIR in mixtures MA/Py 70:30 by weight containing 1 wt.% BDMA, MPAP or CQ/EDMB in combination with 2 wt.% IODS (hollow symbols). The conversion of pure MA using the same photoinitiators (filled symbols) is also shown.

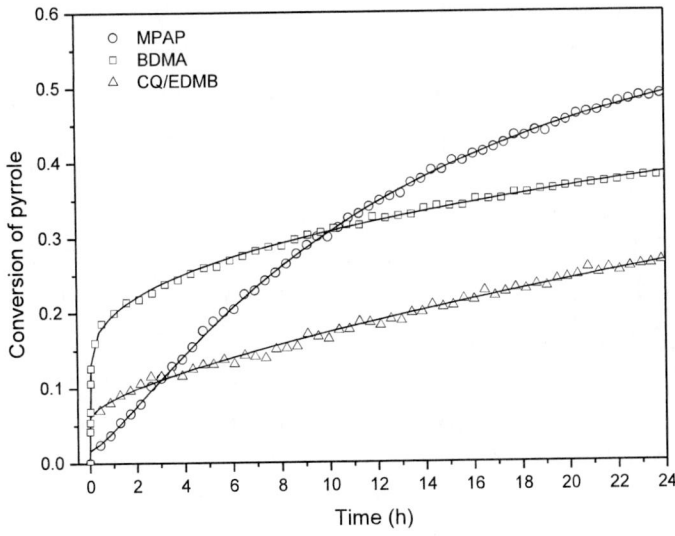

Figure 8. Degree of conversion of pyrrole calculated by MIR spectroscopy in mixtures MA/Py 70:30 by weight containing 1 wt.% of BDMA, MPAP or CQ/EDMB in combination with 2 wt.% IODS.

Free radical photopolymerization of methacrylate monomers is a diffusion controlled process due to the almost instantaneous build-up of high molecular weight and cross-linked polymer [30]. During polymerization, the conversion of monomer to polymer results in an increase in viscosity. This increase, in turn, causes a decrease in both the translational diffusion of monomer and polymer and the segmental diffusion of the polymer. When the diffusional limitations become large enough to restrict the diffusion of growing polymer chains, the termination rate by combination or disproportionation will decrease, causing a buildup in radical concentration and, hence, autoacceleration. Eventually, at greater double bond conversion, the propagation mechanism also becomes diffusion controlled and autodeceleration is observed, as denoted by the rapid decrease in the polymerization rate. The presence of pyrrole in the mixture has two antagonistic effects. The first is a decrease in the concentration of methacrylate groups in the mixture. In this case, a decay of reaction rate should be observed. The second effect is an enhancement of the mobility of the reactive species, pyrrole acting as a solvent and then as a plasticizer during polymerization of the MA methacrylate monomer [30]. According to experimental results (Figure 7), solubilizing and plasticizing effects are more important than the concentration one. On the other hand, Figure 7 also shows that when the IODS salt was combined with the CQ/EDMB pair the conversion of MA in the mixture Py/MA was significantly lower than that in pure MA. This observation can be explained in terms of the mechanism of photolysis of the CQ/EDMB couple [28]. Under visible light irradiation the CQ is excited to form the excited singlet state which is converted to the triplet state CQ* through intersystem crossing. The reaction of CQ* with hydrogen donors generates free radicals by electron and proton transfer through a short lived intermediate complex. As a result donor derived radicals and pinacol are produced. It is well known that the donor-derived radicals initiate the polymerization reaction while radicals derived from the CQ dimerize and are not effective initiators. In addition, if monomers containing hydrogen donor groups such as methylene ether ($-O-CH_2-$) are present while CQ is photoreduced, then hydrogen abstraction occur from the monomer [28].

Thus, it is possible that in formulations containing pyrrole, which has hydrogen donor groups, hydrogen abstraction from EDMB by the CQ is competitively involved with the reaction with Py. This will result in a decreased efficiency to form EDMB derived initiating radicals. The degree of conversion of pyrrole versus time in mixtures MA/Py 70/30 wt.% photoactivated with different photoinitiator systems is shown in Figure 8. The resins were firstly irradiated for 40 s and then the progress of the polymerization was monitored in the absence of radiation, i.e., in the dark. Figure 8 shows that, irrespective of the photoinitiator system used, the extent of conversion reached after 40 s irradiation was bellow 15%. This low extent of polymerization of Py can be attributed to both an inherent low reactivity of the Py monomer and the existence of a diffusion-controlled polymerization reaction. It is conceivable that the formation of the methacrylate network result in a delay in the polymerization of Py. The faster formation of the methacrylate network produces the vitrification of the whole system thereby reducing the molecular mobility of the pyrrole and, consequently, a slow polymerization of pyrrole in the mixture. Nevertheless, the polymerization of pyrrole continues slowly in the dark due to the living nature of cationic polymerization.

Figure 3 illustrates the mechanism of the polymerization of pyrrole. The propagation reaction is produced by the combination of two radical cation monomers or oligomers and the loss of two hydrogen ions [9]. The resultant bond is at the 2 position of the pyrrole ring; thereby forming 2,2'-bipyrrole. 2-monosubstituted pyrroles only form dimmers and 2,5-disubstituted pyrroles do not polymerize. The re-oxidation of the bipyrrole and further combination of radicals continue the propagation reaction. When no monomer is present for oxidative polymerization or side reactions terminate the PPy chain termination occurs. The resultant polymer contains a cationic polymer backbone accompanied by counterions which maintain charge neutrality. These counterions, commonly termed or "dopants" are typically the anions of the chemical oxidant used in the polymerization or reduced product of oxidant. For example, when Cl_2 or $FeCl_3$ are used as oxidants for the polymerization then Cl^- ion is incorporated as counterion into the polymer network. In addition, some dopants are commonly incorporated into the formulation in order to control the electrical properties of the resultant cured material [31-32]. In our research, the counterion is the anion of the IODS; i.e.; SbF_6^-.

BDMA and MPAP have been used as UV radical photoinitiators for many years because they result in a very fast photoinduced polymerization reaction. The different photoinitiation efficacy exhibited by BDMA and MPAP is associated to structural effects on the recombination rate of radicals and to the radical's reactivity toward the double bond. Figure 9 illustrates that irradiation of MPAP at 365 nm results in its cleavage to give benzoyl and methoxybenzyl primary radicals. Equally, irradiation of BDMA at 365 nm generates benzoyl and dimethoxybenzyl radicals followed by a cleavage of the dimethoxybenzyl radicals to give methyl radicals and methyl benzoate. It is worth noting that the quantum yield of BDMA consumption in benzene is nearly unity [33], while the value for MPAP is about 0.24 [33]. Moreover, the molar absorption coefficients of BDMA and MPAP at 365 nm are 170 and 110 (l/mol cm) respectively. From these values, it expected that BDMA result in a higher amount of initiating radicals compared with MPAP.

Concerning the cationic photopolymerization, it has gained increasing interest after the discovery of onium salts by Crivello [34]. Onium salts (On^+) containing aromatic groups such as diaryliodonium ($Ar_2I^+X^-$) or triarylsulfonium ($Ar_3S^+X^-$) salts, with non nucleophilic counterions such as AsF_6^-, SbF_6^-, BF^-, and PF_6^- are highly photosensitive and efficient cationic photoinitiators which can be used as cationic photoinitiators. Irradiation of onium salts with UV light produces a variety of reactive radical, radical-cation and cation intermediates described as follows:

$$Ar_2I^+ \xrightarrow{h\nu} ArI^{+\bullet} + Ar^\bullet \qquad (1)$$

$$ArI^{+\bullet} + RH \rightarrow ArI^+ - H + R^\bullet \qquad (2)$$

$$ArI^+ - H \rightarrow ArH + H^+ \qquad (3)$$

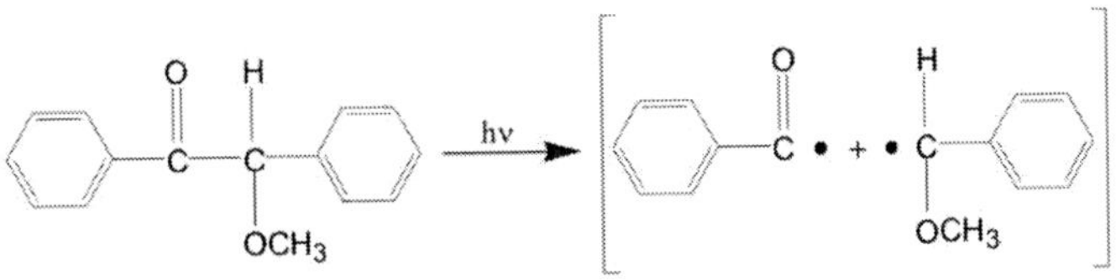

Figure 9. Photolysis of (a) MPAP and (b) BDMA under irradiation at 365 nm.

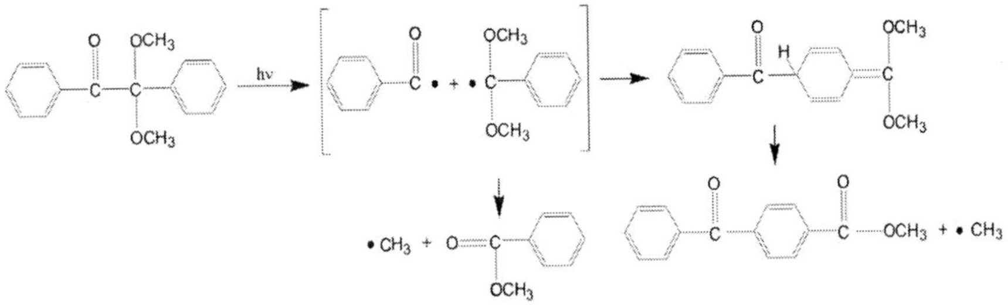

The cationic species generated under irradiation interact with proton donor molecules (RH, i.e., monomer or impurity) to produce the strong Brønsted acid, H^+X^- (Eq. (3)) which is an efficient initiator of cationic polymerization. Moreover, free radicals produced during irradiation of MPAP, BDMA and CQ/EDMB are able reduce the iodonium salt to generate the corresponding carbocations which initiate cationic polymerization as illustrated in Figure 10 [35-38]. In this manner, the presence of α-cleavage type photoinitiators such MPAP and BDMA enables the photoinduced polymerization of both pyrrole and methacrylate groups in a one-pot, one-step procedure. It is worth mentioning that the comparatively higher number of initiating radicals produced during photolysis of BDMA will result in a higher number of polymerizing oligomers compared with those produced by photolysis of MPAP. Consequently, the oligomers produced in resins photoactivated with BDMA will have lower molar mass compared with those produced in resins photoactivated with MPAP. This feature explains the red-shift of the UV-vis spectra in Figures 4 a-b, which occurs with the increase of the size of the oligomers formed [26].

(a)

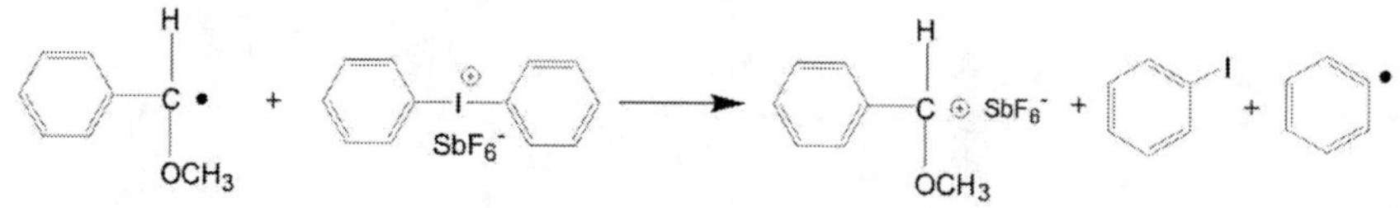

(b)

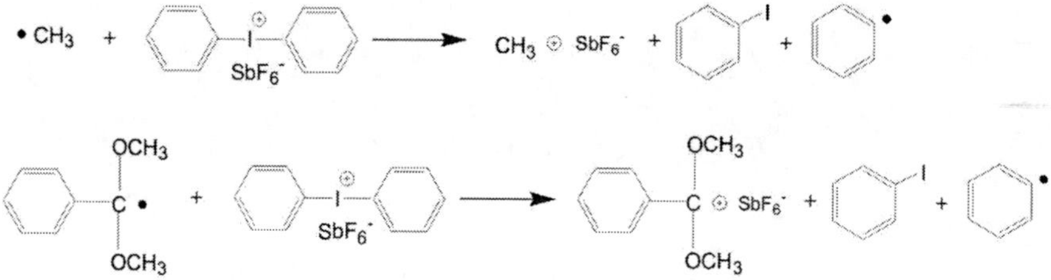

Figure 10. The photochemically generated electron-donating free radicals derived from (a) MPAP and (b) BDMA reduce the iodonium salt to yield carbocations responsible of initiating polymerization of pyrrole.

Visual examination of the cured Py/MA resins showed that polypyrrole was well dispersed in the methacrylate polymer and no phase separation occurred during polymerization. The morphology of the polymers based on Py/MA mixtures were examined by SEM microscopy. Fracture surfaces were examined by scanning electron microscopy (SEM) using a Jeol JSM 35 CF apparatus, after coating the broken surfaces with a thin gold layer. Figure 11 shows that phase separation did not occur and the polypyrrole was well dispersed in the methacrylate polymer in the absence of modifiers. From SEM studies it is concluded that the combination of pyrrole and MA resulted in the formation of interpenetrating polymer networks.

Results from FTIR studies clearly demonstrate that the use of the IODS along with α-cleavage type photoinitiators such as MPAP and BDMA makes possible the simultaneous photopolymerization of pyrrole and methacrylate monomers in a one-pot, one-step procedure. In this manner, 2-3 mm thick layers of interpenetrating networks can be prepared.

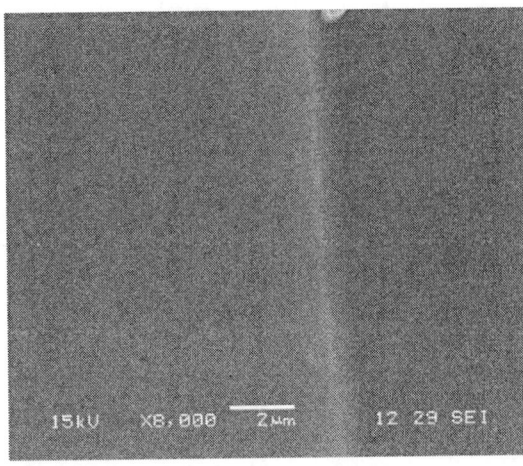

Figure 11. SEM micrograph showing a fractured surface of a 2 mm thick specimen of a mixture Py/MA 50/50 by weight.

Electrical Conductivity

For the electrical measurements, silver electrodes were painted on both faces of the sample specimens. Electrical conductivity (σ_d) of mixtures pyrrole/methacrylate was measured by a two-probe method and expressed as the specific volume resistivity (Ω cm), employing a Super Megohmmeter Hioki DSM-8104, DSM-8542. The values of electrical conductivity, σ_d were calculated from the following equation:

$$\sigma_d = \left(\frac{I}{V}\right)\frac{t}{A} \; (Siemens/m) \tag{4}$$

where t and A are the thickness and area of the specimen, I the current supplied and ΔV is the measured potential drop. Electrical conductivity was also measured in the frequency range 0.10 Hz to 10 MHz, employing both Hioki 3535 and 3522-50 LCR meters.

As illustrated in Figure 3 PPy exhibits resonance structures that bear a resemblance to the aromatic forms. PPy is a no conducting polymer in this neutral state and turns into conducting when it is oxidized. The charge related with the oxidized state is delocalized over several pyrrole units and is able to form a cation radical termed polaron or a dication termed bipolaron. The intrinsic disorder in the structure of polypyrrole has make difficult to determine conclusively the mechanism of electrical conduction in PPy. The mechanism most extensively accepted comprises charge transport along the polymer chains, along with hopping of charge carriers; i.e., holes and bipolarons [10]. The charge on the polymer is neutralized by an anion counterion which is not very mobile within the PPy network. Figure 12 shows the experimental curves I vs V used to calculate the electrical conductivity from Eq (4). The composition of sample specimens used in electrical characterizations and the measured values of electrical conductivity are presented Table 1.

Table 1. Composition of the resins used to prepare sample specimens for measurement of electric conductivity. Py, PPy and IODS are wt.%. All resins contained 1 wt.% MPAP. X_{Py} is the conversion of Py after 24 h reaction determined by ATR. The amount of PPy was calculated from the degree of conversion of Py

Sample	Py	IODS	X_{Py}	Py	σ_d (S/m)
M1	0	-	-	-	1,7 10^{-5}
M2	40	2	0.65	26	2,1 10^{-4}
M3	40	5	0.65	26	0,0014
M4	40	7	0.65	26	0,0041
M5	50	10	0.76	38	0,0335

Values from the experimental curves I versus V presented in Figure 12 were used in Eq (4) to calculate the electrical conductivity. The mixture containing 50 wt.% Py in combination with 10 wt.% IODS and 1 wt.% MPAP displayed significant increase in the electrical conductivity. This marked increase in conductivity is explained in terms of the development of an electrical conducting polymer network within the insulating methacrylate matrix. The mixture containing 50 wt.% Py reached a conversion of Py equal to 0.75 which is equivalent to 36 wt.% PPy. As previously described, the vitrification of the methacrylate network reduces the mobility of the reacting medium and, in turn, affects the polymerization of Py. Thus, mixtures with lower amounts of methacrylate resin are expected to render an enhanced mobility of the reactive species; thereby increasing the conversion of Py (see Table 1). Unfortunately, resins prepared with more than 50 wt.% pyrrole produced brittle polymers having poor mechanical properties. This feature established a limit in the proportion of Py to be used in the Py/MA mixtures.

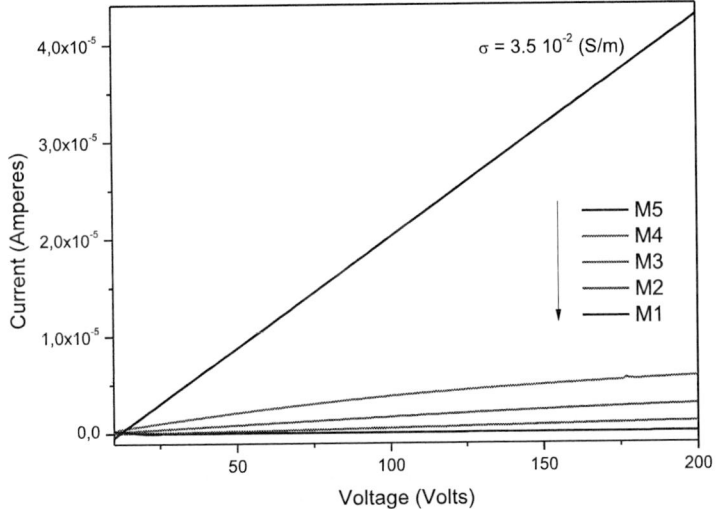

Figure 12. Current vs. voltage in measured in test specimens having the amounts of PPy and photoinitiator system shown in Table 1.

The electrical conductivity of the MA-PPy composites were also analyzed in the frequency range from 10^2 to 10^7 Hz. The frequency dependent electrical conductivities (σ_a) of MA/PPy formulations containing the proportions of PPy and photoinitiator given in Table 1 are depicted in Figure 13.

Once again, the significant increase in the electrical conductivity of the mixture prepared with 50 wt.% Py and 10 wt.% IODS in combination with 1 wt.% MPAP is explained in terms of a comparatively higher proportion of conducting PPy. Unfortunately, mixtures photoactivated with BDMA displayed a more modest increase in conductivity in comparison with that observed in the mixtures photoactivated with MPAP. A possible explanation is that the electronic conductivity is impeded by the lower size of the oligomers resulting from BDMA. At this point, it is worth mentioning that the value of σ_a of the system containing 50 wt.% PPy in combination with 10 wt.% IODS and 1 wt.% MPAP at 10 MHz is comparable to those found for a polyacrylonitrile matrix modified with PPy derivatives [39] and poly N-vinyl carbazole-Fe_3O_4 nanocomposites containing PPy [40].

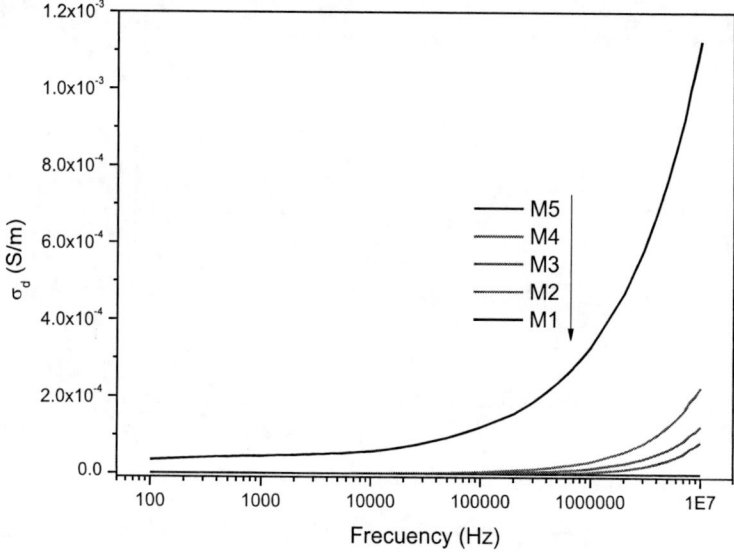

Figure 13. Electric conductivity versus frequency in test specimens having the composition shown in Table 1.

As illustrated in Figures 12-13, electrical conductivity values showed an abrupt increase at a content of Py equal to 50 wt.. %. This feature may be explained according to percolation theory, which proposes that a complete conducting path for the flow of electrical current is generated at a particular amount of conducting polymer. It is well established that when PPy conducting particles are dispersed within an insulating polymer matrix, the composite material exhibits electrical conductivity when the particle content is above a certain value referred to as the percolation threshold [41]. Below the percolation limit the electrical conductivity of the composite material is nearly the same as that of the non conducting polymer because a full conducting path has not been formed. In the vicinity of the percolation limit, a full network for electrons transport is formed because of the contact of the isolated conductive particles. Then, immediately after the percolation limit, a minor increase in the amount of conductive particles may increase markedly the connections in the conducting network [41]. In this manner, a non conducting composite is converted into a conducting material in an abrupt way. Composite materials prepared from conjugated conducting polymers and non-conducting polymers are another class of composite materials that frequently exhibit a percolation threshold [42-44].

In summary, even though mechanistic details still remain to be elucidated, it is clear that Py polymerizes efficiently in mixtures with methacrylate monomers photoactivated with an iodonium salt and free radical photoinitiators. The electrical conductivity of mixtures containing 50 wt.% Py and 10 wt.% IODS in combination with 1 wt.% MPAP encourage further research concerning the use of greater amounts of Py in different photopolymerizable monomers.

Conclusion

Mixtures pyrrole/methacrylate photoactivated with an iodonium salt in combination with BDMA or MPAP are polymerized efficiently by light irradiation at 365-nm. On the other hand, a very slow polymerization of both pyrrole and methacrylate occurred in resins photoactivated with iodonium salt in combination with the CQ/EDMB pair and irradiated at 470 nm.

Pyrrole/methacrylate blends form an interpenetrating polymer network characterized by the absence of phase separation as revealed by microscopy studies.

The electrical conductivity of the polymer networks based on Py/MA was markedly influenced by the proportion of PPy in the formulations.

Mixtures Py Mixtures Py/MA containing up to 50 wt.% Py and photoactivated with 10 wt.% iodonium salt could be prepared without jeopardizing the mechanical properties of the resultant polymers.

ACKNOWLEDGMENTS

The financial support provided by the ANPCyT through PICT 1008 (2015) and CONICET is gratefully acknowledged.

REFERENCES

[1] Biallozor, S, A Kupniewska. Conducting polymers electrodeposited on active metals. *Synthetic Metals* 155 (2005) 443.
[2] Qiu, L, S Zhang, L Zhang, M Sun, W Wang. Preparation and enhanced electrochemical properties of nano-sulfur/poly(pyrrole-co-aniline) cathode material for lithium/sulfur batteries. *Electrochimica Acta* 55 (2010) 4632.
[3] Meyer, H, RJ Nichols, D Schroer, L Stamp .The use of conducting polymers and colloids in the through hole plating of printed circuit boards. *Electrochim Acta* 39 (1994)1325.
[4] Omastová, M, P Bober, Z Morávková, Nikola Peřinka, M Kaplanová, T Syrový, J Hromádková, M Trchová, J Stejskal. Towards conducting inks: Polypyrrole–silver colloids. *Electrochimica Acta* 122 (2014) 296.
[5] Suryawanshia, AJ, P Thuptimdang, J Byroma, E Khanb, VJ Gelling. Template free method for the synthesis of Ag–PPy core–shell nanospheres with inherent colloidal stability. *Synthetic Metals* 197 (2014) 134.
[6] Inzelt, G, *Conducting polymers:a new era in electrochemistry*, Chapter 4. Springer, Berlin (2008).
[7] Freund, S, *BA Deore Self-doped conducting polymers*, Chapter 1. Wiley, Hoboken (2007).

[8] Chougule, MA, SG Pawar, PR Godsa, RN Mulia, S Seb, VB Patil. Synthesis and Characterization of Polypyrrole (PPy) Thin Films. *Soft Nanoscience Letters.* 1 (2011) 6.

[9] Tan, Y, K Ghandi. Kinetics and mechanism of pyrrole chemical polymerization. *Synthetic Metals* 175 (2013) 183.

[10] Ansari, R, Polypyrrole Conducting Electroactive Polymers: Synthesis and Stability Studies. *E-Journal of Chemistry.* 3 (2006) 186.

[11] Singh, RN, R Awasthi. *Polypyrrole composites: electrochemical synthesis, characterizations and applications, electropolymerization,* Dr. Ewa Schab-Balcerzak (Ed.) (2011).

[12] Sangian, D, W Zheng, GM Spinks. Optimization of the sequential polymerization synthesis method forpolypyrrole films. *Synthetic Metals* 189 (2014) 53.

[13] Campomanes, R, SE Bittencourt, JSC Campos. Study of conductivity of polypyrrol-poly(vinyl alcohol) composites obtained photochemically. *Synthetic Metals* 102 (1999) 1230.

[14] Han, G, G Shi. Porous polypyrrole/polymethylmethacrylate composite film prepared by vapor deposition polymerization of pyrrole and its application for ammonia detection. *Thin Solid Films* 515 (2007) 6986.

[15] Rinaldi, AW, MH Kunita, MJ Santos, E Radovanovic, AF Rubira, EM Girotto. Solid phase photopolymerization of pyrrole in poly(vinylchloride) matrix. *Europ Polym J* 41 (2005) 2711.

[16] Takano, T, A Mikazuki, T Kobayashi. Conductive polypyrrole composite films prepared using wet cast technique with a pyrrole–cellulose acetate solution. *Polym Eng & Sci.* 54, (2014) 78.

[17] Migahed, MD, T Fahmy, M Ishra, A Barakat. Preparation, characterization, and electrical conductivity of polypyrrole composite films. *Polymer Testing* 23 (2004) 361.

[18] Rabek, JF, J Luck, M Zuber, BJ Que, WF Shie. Photopolymerization of pyrrole initiated by the ferrocene- and iron-arene salts- chlorinated solvents complexes. *J Macromol Scie* A 29 (1992) 297.

[19] Kobayashi, N, K Teshima, R Hirohashia. Conducting polymer image formation with photoinduced electron transfer reaction. *J Mater Chem.* 8 (1998) 497.

[20] Martins, CR, M Azevedo. Metal nanoparticles incorporation during the photo polymerization of polypyrrole. *J Mater Sci* 41 (2006) 7413.

[21] Hodko, D, M Gamboa-Aldeco, O Murphy. Photopolymerized silver-containing conducting polymer films. Part I. An electronic conductivity and cyclic voltammetric investigation. *J Solid State Electrochem* 13 (2009) 1063.

[22] Hodko, D, M Gamboa-Aldeco, OJ Murphy. Photopolymerized silver-containing conducting polymer films. Part II. Physico-chemical characterization and mechanism of photo polymerization process. *J Solid State Electrochem* 13 (2009) 1077.

[23] Kasisomayajula, SV, X Qi, C Vetter, K Croes, D Pavlacky, VJ Gelling. A structural and morphological comparative study between chemically synthesized and photopolymerized poly(pyrrole*). J. Coat. Technol. Res.,* 7 (2010) 145.

[24] Cai, Y, JLP Jessop. Decreased oxygen inhibition in photopolymerized acrylate/epoxide hybrid polymer coatings as demonstrated by Raman spectroscopy. *Polymer* 47 (2006) 6560.

[25] Jin, S, X Liu, W Zhang, Y Lu, G Xue. Electrochemical copolymerization of pyrrole and styrene. *Macromolecules* 33 (2000) 4805.

[26] Bae, WJ, KH Kim, HW Jo. A water-soluble and self-doped conducting polypyrrole graft copolymer. *Macromolecules* 38 (2005)1044.

[27] Bredas, JL, JC Scott, K Yakushi, B Street. Polarons and bipolarons in polypyrrole: Evolution of the band structure and optical spectrum upon doping. *Phys Rev* B 30 (1984) 1023.

[28] Asmussen, S, G Arenas, W Cook, C Vallo. Photobleaching of camphorquinone during polymerization of dimethacrylate-based resins. *Dent Mater* 25 (2009) 1603.

[29] Workman, J, L Weyer. *Practical guide to interpretative Near-Infrared spectroscopy.* Boca Raton: CRC Press (2008).

[30] Asmussen, S, W Schroeder, I dell'Erba, C Vallo. Monitoring of visible light photo polymerization of an epoxy/dimethacrylate hybrid system by Raman and near-infrared spectroscopies. *Polymer Testing* 32 (2013) 1283.

[31] Wen, Q, X Pan, Q Hu, S Zhao, Z Hou, Q Yu. Structure–property relationship of dodecylbenzenesulfonic acid doped polypyrrole. *Synthetic Metals* 164 (2013) 27.

[32] Omastová, M, M Trchová, J Kovářová, J Stejskal. Synthesis and structural study of polypyrroles prepared in the presence of surfactants. *Synthetic Metals* 138 (2003) 447.

[33] Pokhrel, MR, K Janik, SH Bossmann. Photoinitiated synthesis and characterization of P(MMA/DPB) polymer. *Macromolecules* 33 (2000) 3577.

[34] Crivello, JV, The discovery and development of onium salt cationic photoinitiators. *J Polym Sci:A: Polym Chem.* 37 (1997) 4241.

[35] Tehfe, M-Ali, J Lalevée, F Morlet-Savary, N Blanchard, C Fries, B Graff, X Allonas, F Louërat, JP Fouassier. Near UV-visible light induced cationic photopolymerization reactions: A three component photoinitiating system based on acridinedione/silane/iodonium salt. *Europ Polym J,* 46 (2010) 2138.

[36] Degirmenci, M, Y Hepuzer, Y Yagci. One-step, one-pot photoinitiation of free radical and free radical promoted cationic polymerizations. *J Appl Polym Sci,* 85 (2002) 2389.

[37] Xiao, P, J Zhang, F Dumur, MA Tehfe, F Morlet-Savary, B Graff, D Gigmes, JP Fouassier, J Lalevée. Visible light sensitive photoinitiating systems: Recent progress in cationic and radical photopolymerization reactions under soft conditions. *Prog Polym Sci* 41 (2015) 32.

[38] *Photoinitiators for Polymer Synthesis: Scope, Reactivity, and Efficiency.* Fouassier, JP, Jacques Lalevée (Editors), Wiley-VCH, New York (2012).

[39] Cetiner, S, H Karakaa, R Ciobanu, M Olariu, NU Kaya, C Unsal, F Kalaoglu, ASi Sarac. Polymerization of pyrrole derivatives on polyacrylonitrile matrix, FTIR–ATR and dielectric spectroscopic characterization of composite thin films. *Synthetic Metals* 160 (2010) 1189.

[40] Haldar, I, M Biswas, A Nayak. Microstructure, dielectric response and electrical properties of polypyrrole modified (poly N-vinyl carbazole–Fe_3O_4) nanocomposites. *Synthetic Metals* 161 (2011) 1400.

[41] Omastova, M, I Chodak, J Pionteck. Electrical and mechanical properties of conducting polymer composites. *Synthetic Metals* 102 (1999) 1251.

[42] Omastova, M, S Kogna, J Pionteck, A Janke, J Pavlinec. Electrical properties and stability of polypyrrole containing conducting polymer composites. *Synthetic Metals* 81 (1996) 49.

[43] Fournier, J, G Boiteux, G Seytrea, G Marichy. Percolation network of polypyrrole in conducting polymer composites. *Synthetic Metals* 84 (1997) 839.

[44] Mandal, TK, BM Mandal. Interpenetrating polymer network composites of polypyrrole and poly (methyl acrylate) or poly (styrene-co-butyl acrylate) with low percolation thresholds. *Synthetic Metals* 80 (1996) 83.

INDEX

A

absorption spectra, 74, 75
acetic acid, 52, 62
acetophenone, 57
acid, viii, 2, 5, 11, 24, 25, 36, 42, 48, 49, 50, 52, 54, 56, 57, 59, 60, 61, 62, 64, 87, 97
acrylate, 96, 98
actuation, 32
adaptability, 48
additives, ix, 53, 62, 72, 73
adhesion, 9, 10, 14, 15, 40, 41
adhesion strength, 14
adhesives, 7
adsorption, 10, 14
aesthetic, viii, 2, 3
aliphatic amines, 59, 61
alkaloids, viii, 45, 46
amine group, 59
amino acids, 48, 50
ammonia, 33, 35, 42, 95
ammonium, 5, 9, 37, 49, 59
aniline, 56, 61, 62, 65, 94
anisotropy, 17
antioxidant, 46, 47

aqueous phase, v, ix, 45, 46, 48, 52, 56, 59, 60, 61, 62, 65
atmospheric ageing, 2, 9, 25
atmospheric aging, viii, 2
atmospheric pressure, 9, 15, 41
autodeceleration, 84

B

bonding, 3, 4, 17, 26, 27, 29, 30, 31, 62
bonds, 3, 27, 51, 64

C

catalyst, 49, 50, 56, 58, 60, 61, 64, 65
catalytic activity, 58
cation, 4, 6, 12, 28, 50, 85, 86, 90
C-C, 26, 29, 62, 64
cellulose, 17, 73, 95
chain propagation, 6, 27
chemical, vii, viii, ix, 2, 4, 5, 7, 8, 10, 12, 13, 14, 16, 22, 24, 26, 29, 30, 31, 35, 36, 37, 38, 42, 48, 56, 71, 72, 73, 74, 85, 95, 96

Index

chemical characteristics, 38
chemical degradation, 24
chemical reactions, 7, 48
chemical stability, 4
chemicals, 46, 47
cholesterol, 46
coatings, vii, 1, 7, 11, 19, 21, 22, 39, 72, 96
compounds, 34, 46, 47, 48, 49, 56, 57, 59
computational capacity, 32
condensation, 48, 49, 56, 60, 61, 64
conducting polymer composites, 34, 41, 98
conductive textiles, vii, 1, 2, 36, 40
conductivity, vii, viii, x, 1, 2, 3, 4, 5, 6, 9, 10, 11, 13, 14, 17, 21, 24, 25, 26, 27, 30, 31, 33, 34, 35, 36, 37, 40, 41, 42, 72, 73, 74, 89, 90, 91, 92, 93, 95, 96
condudting polymers, 72
cotton, 10, 11, 14, 15, 16, 17, 18, 19, 22, 24, 25, 36, 37, 38
covalent bonding, 51
cyclization, 48, 50, 52, 54, 57
cyclodextrins, 61, 65

D

deposition, 8, 9, 10, 11, 13, 16, 20, 21, 28, 31, 36, 39, 95
derivatives, vii, viii, 45, 46, 48, 49, 50, 51, 52, 53, 55, 58, 59, 60, 61, 62, 64, 65, 66, 72, 92, 98
diffraction, viii, 2, 9, 17, 19
diversified, 48, 52, 59
dopants, viii, 2, 17, 20, 21, 22, 24, 25, 85
doping, 9, 11, 27, 29, 34, 96
double bonds, 3, 34

E

eco-friendly, ix, 46, 48, 49, 50, 52, 54, 55, 56, 58, 59, 62, 66
electric conductivity, viii, 2, 3, 90

electrical characterization, 90
electrical conductivity, vii, x, 2, 4, 13, 17, 20, 26, 31, 36, 72, 74, 89, 90, 91, 92, 93, 95
electrical properties, vii, x, 72, 86, 98
electromagnetic, vii, 1, 2, 36, 40
electron, x, 7, 13, 24, 27, 28, 34, 43, 56, 63, 64, 72, 84, 88, 96
electron microscopy, x, 72

F

films, 12, 25, 30, 33, 39, 40, 41, 42, 73, 81, 95
formation, vii, viii, x, 2, 4, 6, 7, 21, 24, 27, 28, 30, 31, 51, 57, 58, 61, 62, 65, 72, 76, 77, 80, 85, 88, 96
free radicals, 84, 87, 88
FTIR, viii, 2, 9, 13, 16, 17, 22, 23, 26, 29, 30, 31, 36, 74, 80, 81, 88, 98
FTIR spectroscopy, 80

G

gem-diactivated, 52, 64
glucose oxidase, 33, 42
glycol, 55
growing polymer chain, 84

H

halogen(s), 5, 38
hybrid, vii, x, 37, 72, 96, 97
hydrazine, 59
hydrogen, 6, 7, 28, 31, 62, 84, 85
hydrogen abstraction, 7, 84
hydrogen peroxide, 31
hydroxyl, 57, 58, 59, 62
hydroxyl groups, 57, 62

Index

I

infrared spectroscopy, 81
interactive textiles, 2
interfacial bonding, 9
interference, vii, 1, 2
intermediate, 48, 50, 54, 59, 64, 84
iodonium salt, ix, 72, 73, 74, 87, 88, 93, 94, 97
ionic liquid, 48, 50
ions, 5, 6, 14, 17, 24, 25, 26, 27, 35, 36, 85
IR spectra, 58
irradiation, 6, 12, 26, 27, 28, 30, 31, 35, 39, 55, 76, 78, 79, 81, 82, 84, 86, 87, 93

J

joint pain, 33

K

ketones, 61

L

LED, 74, 76, 78, 79, 80
light, x, 7, 8, 72, 73, 74, 77, 80, 81, 84, 93, 97
lying, 17

M

material sciences, vii, viii, 45
materials, vii, 1, 2, 7, 9, 11, 34, 36, 74, 93
mechanical properties, ix, 71, 72, 73, 91, 94, 98
mechanism, 4, 7, 26, 27, 28, 34, 35, 49, 50, 52, 53, 54, 56, 57, 58, 59, 63, 65, 73, 76, 78, 80, 84, 85, 90, 95, 96
metals, 9, 54, 94
methacrylate, v, vii, ix, 71, 72, 73, 74, 76, 81, 82, 84, 87, 88, 89, 90, 93
microelectronics, 7
microfabrication, 7
microscopy, 88, 93
microwave radiation, 37
molecules, vii, viii, 11, 21, 24, 26, 27, 34, 35, 45, 46, 49, 56, 69, 87
monomer molecules, 20
monomers, 5, 20, 73, 74, 75, 81, 84, 85, 88, 93
morphology, viii, 2, 8, 9, 13, 21, 28, 30, 36, 88
multi-component, 51

N

nanocomposites, 40, 92, 98
nanoparticles, 60, 72, 96
nanotube, 38
naphthalene, 5
neutral, 51, 61, 62, 89
nitrogen, 3, 35, 50, 52
NMR, 25, 41, 62, 65
NSA, 5, 21, 23, 24, 31
nucleophilicity, 63

O

olefins, 52
oligomers, 61, 77, 85, 87, 92
organic compounds, ix, 12, 46
organic solvents, 47, 50, 59
oxidation, 4, 5, 6, 22, 24, 26, 72, 85
oxygen, 24, 96

P

paradigm shift, ix, 46, 48
percolation, 92, 98

percolation theory, 92
photopolymerization, ix, 71, 72, 73, 74, 84, 86, 88, 95, 97
poly(vinylchloride), 73, 95
polymer, vii, ix, 3, 4, 5, 6, 7, 8, 12, 16, 17, 20, 21, 25, 27, 34, 38, 39, 41, 71, 72, 81, 84, 85, 88, 89, 90, 92, 93, 96, 97, 98
polymer blends, 12
polymer chain(s), 3, 6, 12, 17, 27, 34, 90
polymer clusters, 34
polymer films, 81, 96
polymer matrix, 5, 92
polymer networks, 12, 88, 93
polymeric materials, 6
polymerization, vii, viii, ix, 2, 4, 5, 6, 7, 8, 10, 11, 12, 14, 16, 18, 19, 20, 21, 22, 23, 25, 26, 27, 28, 29, 30, 31, 33, 35, 36, 37, 38, 41, 71, 73, 74, 76, 78, 79, 80, 81, 82, 84, 85, 86, 87, 88, 90, 93, 95, 96, 97
polymerization mechanism, 80
polymerization process, ix, 8, 28, 71, 96
polymerization processes, ix, 8, 71
polymerization time, 37
polymers, vii, viii, ix, 2, 3, 8, 17, 34, 39, 41, 42, 71, 72, 73, 88, 91, 93, 94
polymethylmethacrylate, 73, 95
polypyrrole, v, vii, viii, ix, 1, 2, 3, 4, 5, 6, 9, 10, 11, 12, 14, 15, 17, 18, 19, 24, 25, 27, 29, 30, 31, 33, 35, 36, 37, 38, 39, 40, 41, 42, 43, 71, 72, 73, 75, 76, 80, 88, 90, 94, 95, 96, 97, 98
pyrrole(s), v, vii, viii, ix, 3, 4, 5, 6, 9, 10, 11, 12, 13, 17, 18, 19, 20, 22, 26, 27, 28, 29, 30, 33, 37, 41, 45, 46, 47, 48, 49, 50, 51, 52, 53, 54, 55, 56, 57, 58, 59, 60, 61, 62, 63, 64, 65, 66, 71, 72, 73, 74, 76, 78, 80, 81, 82, 83, 84, 85, 87, 88, 89, 91, 93, 94, 95, 96, 98

Q

quartz, 75

R

radiation, 7, 11, 12, 13, 26, 27, 39, 77, 80, 85
radical mechanism, 7, 73, 76
radical polymerization, 39
radicals, 7, 12, 27, 84, 85, 86, 87
Raman spectroscopy, 81, 96
reactants, 49, 52, 58
reaction mechanism, 52, 59, 65
reaction rate, 84
reaction time, 53, 58
reactions, ix, 5, 6, 7, 27, 46, 48, 50, 51, 56, 57, 62, 64, 85, 97
reactivity, 56, 62, 63, 85, 86
reagents, 51, 54
resins, ix, 72, 73, 74, 77, 80, 81, 85, 87, 88, 90, 91, 93, 96
reusable, 48, 52, 56
rubber, 9

S

salts, 9, 86, 95
scanning electron microscopy, 88
SEM micrographs, 16
semiconductor, 3, 38, 74
sensing, 32, 33, 34, 42
sensitivity, 33, 34
sensor(s0, 32, 33, 34, 42
signal transduction, 34
silver, 72, 89, 94, 96
solubility, ix, 46, 58
solution, 4, 5, 9, 10, 11, 12, 28, 56, 95
solvents, 5, 47, 51, 57, 58, 59, 62, 95
species, 4, 28, 50, 58, 84, 87, 91

spectroscopy, 73, 74, 76, 81, 83, 97
stability, viii, 2, 3, 24, 25, 36, 40, 41, 50, 94, 98
stabilizers, 40
structure, viii, 2, 3, 9, 17, 20, 21, 24, 36, 37, 50, 58, 74, 75, 90, 96
styrene, 96, 98
substitution, 54
synthesis, vii, viii, 2, 3, 5, 9, 11, 16, 21, 22, 26, 35, 39, 45, 46, 47, 49, 51, 52, 54, 56, 57, 58, 59, 60, 62, 64, 65, 66, 72, 94, 95, 97

T

task-specific, 50
tautomerization, 54
techniques, 3, 12, 30, 31, 73, 81
temperature, viii, 2, 5, 8, 10, 12, 25, 26, 33, 34, 37, 39, 41, 42
tensile strength, 13, 19
tetrahydrofuran, 52, 55, 56
textiles, vii, viii, 1, 2, 8, 14, 31, 32, 36, 37, 40, 42
transition metal ions, 5, 38

U

ultrasound, 55, 59
UV irradiation, x, 28, 31, 72
UV light, 86
UV radiation, 12
UV-irradiation, 29, 31

V

vibration, 26, 29, 82
viscosity, 84
volatile organic compounds, 33

W

wavelengths, 7

X

X-ray diffraction (XRD), 13, 43, 17, 18, 60

Related Nova Publications

CRYSTAL VIOLET: PRODUCTION, APPLICATIONS AND PRECAUTIONS

EDITOR: Victor Duffet

SERIES: Chemistry Research and Applications

BOOK DESCRIPTION: *Crystal Violet: Production, Applications and Precautions* opens by presenting the main factors influencing the metachromatic phenomenon. The hypsochromic effect is due to the symmetry of the molecule providing the common electronic signal around the central carbon atom.

SOFTCOVER ISBN: 978-1-53615-806-9
RETAIL PRICE: $82

SUPERCRITICAL CARBON DIOXIDE

AUTHOR: Yizhak Marcus

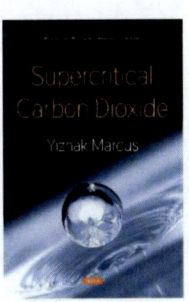

SERIES: Chemistry Research and Applications

BOOK DESCRIPTION: Co-solvents are frequently added to SCD, and the properties of the resulting supercritical fluid mixtures are described in this book. Once either pure SCD or its mixture with a co-solvent have been produced, and the temperature and pressure have been suitably adjusted, the solvent power of this fluid is employed in a great variety of applications that are reviewed and discussed. In particular, the manner of adherence of such fluids to the requirements from 'green' solvents are assessed.

SOFTCOVER ISBN: 978-1-53615-165-7
RETAIL PRICE: $82

To see a complete list of Nova publications, please visit our website at www.novapublishers.com

Related Nova Publications

Understanding Halogenation

Editor: Vladimir T. Phelps

Series: Chemistry Research and Applications

Book Description: *Understanding Halogenation* explores how halogenation stress could participate in neuronal damage of the nervous system and parkinsonian deficits.

Softcover ISBN: 978-1-53615-947-9
Retail Price: $82

Imines: An Overview

Editor: Gordon Sjögren

Series: Chemistry Research and Applications

Book Description: In the opening chapter of *Imines: An Overview*, attempts will be made to provide a brief survey of di-imines derived from benzil hydrazones through imine chemistry. Imines are compounds containing an azomethine linkage C=N connected to hydrogen or carbon atoms.

Softcover ISBN: 978-1-53616-007-9
Retail Price: $95

To see a complete list of Nova publications, please visit our website at www.novapublishers.com